Myths and Science of Soils of the Tropics

Myths and Science of Soils of the Tropics

Proceedings of an international symposium sponsored by Division A-6 of the American Society of Agronomy, the World Association of Soil and Water Conservation, and the Soil and Water Conservation Society, in Las Vegas, Nevada, 17 Oct. 1989.

Editors
R. Lal and P. A. Sanchez

Organizing Committee
R. Lal and T. J. Smith

Editorial Committee
R. Lal, P. A. Sanchez, and T. J. Smith

Editor-in-Chief ASA
G. A. Peterson

Editor-in-Chief SSSA
R. J. Luxmoore

Managing Editor
S. H. Mickelson

SSSA Special Publication Number 29

**Soil Science Society of America, Inc.
American Society of Agronomy, Inc.
Madison, Wisconsin, USA**

1992

Cover Design: Patricia Scullion

Copyright © 1992 by the Soil Science Society of America, Inc.
American Society of Agronomy, Inc.

ALL RIGHTS RESERVED UNDER THE U.S. COPYRIGHT ACT OF 1976 (P.L. 94-553)

Any and all uses beyond the limitations of the "fair use" provision of the law require written permission from the publisher(s) and/or the author(s); not applicable to contributions prepared by officers or employees of the U.S. Government as part of their official duties.

Soil Science Society of America, Inc.
American Society of Agronomy, Inc.
677 South Segoe Road, Madison, WI 53711, USA

Second Printing 1992

Library of Congress Cataloging-in-Publication Data

Myths and science of soils of the tropics: proceedings of an international symposium sponsored by Division A-6 of the American Society of Agronomy, the World Association of Soil and Water Conservation, and the Soil and Water Conservation Society, in Las Vegas, Nevada, 17 Oct. 1989 / editors, R. Lal and P.A. Sanchez; organizing committee, R. Lal and T.J. Smith ... [et al.].
    p.   cm. — (SSSA special publication; no. 29)
  ISBN 0-89118-800-2
  1. Soils—Tropics—Congresses. I. Lal, R. II. Sanchez, P.A. III. American Society of Agronomy. Division A-6. IV. Series.
S599.9.T76M87 1992
631.4'913—dc20   92-4251
                                                              CIP

Printed in the United States of America

CONTENTS

FOREWORD vii
PREFACE ix
CONTRIBUTORS xi
CONVERSION FACTORS FOR SI AND NON-SI UNITS xiii

1 Soil Diversity in the Tropics: Implications for Agricultural Development
  H. Eswaran, F.H. Beinroth, J. Kimble, and T. Cook 1

2 Organic Matter Dynamics in Soils of the Tropics—From Myth to Complex Reality
  D.J. Greenland, A. Wild, and D. Adams 17

3 Myths and Science about the Chemistry and Fertility of Soils in the Tropics
  Pedro A. Sanchez and Terry J. Logan 35

4 Some Aspects of Fertility Associated with the Mineralogy of Highly Weathered Tropical Soils
  U. Schwertmann and A.J. Herbillon 47

5 Soil Physical Properties of the Tropics: Common Beliefs and Management Restraints
  D.K. Cassel and R. Lal 61

6 Relation between Climate and Soil Productivity in the Tropics
  M.V.K. Sivakumar, A. Manu, S.M. Virmani, and E.T. Kanemasu 91

7 Myths and Science of Fertilizer Use in the Tropics
  A. Uzo Mokwunye and L.L. Hammond 121

8 Legume Response to Rhizobial Inoculation in the Tropics: Myths and Realities
  P.W. Singleton, B.B. Bohlool, and P.L. Nakao 135

9 Impact of Soil Fauna on the Properties of Soils in the Humid Tropics
  P. Lavelle, A.V. Spain, E. Blanchart, A. Martin, and S. Martin 157

FOREWORD

Soils of the tropics have many common characteristics and properties with the intensively studied soils of the temperate regions. It is their unique properties, however, that make them worthy of this special publication. A distinguished group of soil scientists from throughout the world has contributed their expertise to bring into one volume much of what is known about the special properties of the soils of the tropics.

This is a most appropriate time for publication of this information. Much attention is currently focused on the ability of soils to sustain the food production and the great biological diversity that is unique to the tropical regions. Although the emphasis is on soil properties related to agricultural production, this volume will be extremely useful to the ecologists, environmentalists, and agriculturalists interested in understanding soil resources and their relationship to climate, flora, and fauna.

William W. McFee, *president*
Soil Science Society of America

Donald N. Duvick, *president*
American Society of Agronomy

PREFACE

Tropical agriculture is faced with a serious challenge of feeding about 70% of world's inhabitants, and meeting other basic necessities of life for 75 to 80% of population of the region that depends on farming. A significant portion of population in tropical countries suffers from malnutrition. In addition to economic issues, intensifications and extension of agriculture to marginal lands have created severe ecological problems (e.g., deforestation, soil degradation, pollution of water and natural environment, and increased greenhouse gas emissions).

A considerable increase in food production in two decades of 1970s and 1980s was achieved by bringing new land under agricultural production. Reserves of the potentially arable prime agricultural land in the tropics are, however, shrinking. Densely populated tropical Asia has few additional land to convert to agriculture. Most potentially, arable land in tropical Africa and South America is located within fragile and ecologically sensitive regions (e.g., tropical rainforest). The potentially productive agricultural land is either inaccessible, too steep, too shallow, or is in regions with too little or too much water, or the essential inputs for crop production are not available.

The apparent imbalance between soil, food, and population growth in the tropics raises several questions. Are soil resources of the tropics capable of sustaining high and economic population? What are the potentials and production constraints of these soils? What are the processes, factors, and causes of soil degradation? How can the degradation trends be reversed? Can high production be achieved without degrading the soil and environment?

There are several misconceptions about soils of the tropics. These misconceptions and myths are based on inadequate information on principal soils of the region, interaction between soils and prevalent climate, soil physical and mineralogical properties, soil chemical and nutritional characteristics, soil biota and their effects on productivity. Myths are propagated by perpetual food crisis, agrarian stagnation, severe problems of soil and environmental degradation and resultant economic and socio-political instability.

It is time that myths regarding soils of the tropics are replaced by scientific realities. We need to strengthen the database so that land capability can be assessed, ecologically compatible soil and crop management systems can be developed and validated, and long-term planning can be made to adopt strategies for sustaining agricultural growth and preserving productive potential of the soil resource. It was these concerns that led A-6 Division of ASA to organize a symposium in 1989 that addressed the issue of "Myths and Science of the Soils of the Tropics."

The symposium was organized with full support and excellent cooperation of Dr. T.J. Smith of the North Carolina State University. As a chair of the A-6 Division, he was the driving force in getting the logistic support for organizing the symposium. We gratefully acknowledge the financial support received from the American Society of Agronomy and the Soil Science Society of America. All authors were extremely cooperative in preparing their manuscript and making necessary revisions. Help received from Sherri Mickelson and other staff of ASA Headquarters in getting the book ready is highly appreciated.

Rattan Lal, *co-editor*
Ohio State University, Columbus, Ohio

Pedro A. Sanchez, *co-editor*
Nairobi, Kenya

CONTRIBUTORS

D. E. Adams	Research Officer, Soil Science Department, University of Reading, London Road, Reading, Berks, RG1 5AQ, U.K.
F. H. Beinroth	Professor of Soil Science, Department of Agronomy and Soils, University of Puerto Rico, Mayaguez, PR 00708
E. Blanchart	Charge de Recherches, Ecole Normale Superieure, Laboratoire d'Ecologie, 46 Rue d'Ulm, 75230 Paris Cedex 05, FRANCE
B. B. Bohlool	(Deceased) Professor of Soil Microbiology, NifTAL Project, University of Hawaii, 1000 Holomua Avenue, Paia, HI 96779-9744
D. K. Cassel	Department of Soil Science, North Carolina State University, Raleigh, NC 27695-7619
T. Cook	Soil Scientist, USDA-SCS, P.O. Box 2890, Washington, DC 20013
H. Eswaran	National Leader of World Soil Resources, USDA-SCS, P.O. Box 2890, Washington, DC 20013
D. J. Greenland	Director, Scientific Services, CAB-International, Wallingford, Oxford, OX10 8DE, U.K.
L. L. Hammond	Director, Agro-Economic Division, International Fertilizer Development Center, Muscle Shoals, AL 35662. Currently Manager, Agronomic Program, Texasgulf Inc., P.O. Box 30321, Raleigh, NC 27622
A. J. Herbillon	Director, Centre de Pedologie Biologique du C.N.R.S., B.P. 5, 54501 Vandoeuvre-les-Nancy Cédex, FRANCE
E. T. Kanemasu	Professor and Head, Department of Agronomy, The University of Georgia, Georgia Station, Griffin, GA 30223-1797
J. M. Kimble	Research Soil Scientist, USDA-SCS, National Soil Survey Center, Federal Building Room 152, 100 Centennial Mall North, Lincoln, NE 68508-3866
R. Lal	Professor of Soil Physics, Agronomy Department, The Ohio State University, Columbus, OH 43210
P. Lavelle	Professor of Ecology, Ecole Normale Superieure, Laboratoire d'Ecologie, 46 Rue d'Ulm, 75230 Paris Cedex 05, FRANCE
Terry J. Logan	Professor of Soil Chemistry, Agronomy Department, The Ohio State University, Columbus, OH 43210
A. Manu	Research Team Leader, Department of Soil and Crop Sciences, Texas A&M University, College Station, TX 77843
A. Martin	Agregee Preparateur, Ecole Normale Superieure, Laboratoire d'Ecologie, 46 Rue d'Ulm, 75230 Paris Cedex 05, FRANCE
S. Martin	Charge de Mission, Ecole Normale Superieure, Laboratoire d'Ecologie, 46 Rue d'Ulm, 75230 Paris Cedex 05, FRANCE

CONTRIBUTORS

A. Uzo Mokwunye — Director, International Fertilizer Development Center, BP 4483, Lome, Togo, AFRICA

P. L. Nakao — Research Associate, NifTAL Project, University of Hawaii, 1000 Holomua Ave., Paia, HI 96779

Pedro A. Sanchez — Professor of Soil Science, North Carolina State University, Raleigh, NC 27695-7619. Currently ICRAF House, Gigiri, Off Limuru Road, Box 30677, Nairobi, Kenya, AFRICA

U. Schwertmann — Professor of Soil Science, Institute of Soil Science, Techn. University of Munich, D-8050 Freising-Weihenstephan, WEST GERMANY

P. W. Singleton — Associate Professor of Agronomy, NifTAL Project, University of Hawaii, 1000 Holomua Avenue, Paia, HI 96779-9744

M. V. K. Sivakumar — Principal Agroclimatologist, ICRISAT Sahelian Center, B.P. 12404, Niamey, Niger (via Paris), AFRICA

A. V. Spain — Senior Scientist, Queensland National Parks and Wildlife Science, P.O. Box 5391, Townsville Mail Centre, Queensland 4810, AUSTRALIA

S. M. Virmani — Principal Agroclimatologist, ICRISAT, Patancheru P.O., 502 324 Andhra Pradesh, INDIA

A. Wild — Professor of Soil Science, Soil Science Department, University of Reading, London Road, Reading, Berks. RG1 5AQ, U.K.

Conversion Factors for SI and non-SI Units

Conversion Factors for SI and non-SI Units

To convert Column 1 into Column 2, multiply by	Column 1 SI Unit	Column 2 non-SI Unit	To convert Column 2 into Column 1, multiply by
Length			
0.621	kilometer, km (10^3 m)	mile, mi	1.609
1.094	meter, m	yard, yd	0.914
3.28	meter, m	foot, ft	0.304
1.0	micrometer, μm (10^{-6} m)	micron, μ	1.0
3.94×10^{-2}	millimeter, mm (10^{-3} m)	inch, in	25.4
10	nanometer, nm (10^{-9} m)	Angstrom, Å	0.1
Area			
2.47	hectare, ha	acre	0.405
247	square kilometer, km^2 (10^3 m)^2	acre	4.05×10^{-3}
0.386	square kilometer, km^2 (10^3 m)^2	square mile, mi^2	2.590
2.47×10^{-4}	square meter, m^2	acre	4.05×10^3
10.76	square meter, m^2	square foot, ft^2	9.29×10^{-2}
1.55×10^{-3}	square millimeter, mm^2 (10^{-3} m)^2	square inch, in^2	645
Volume			
9.73×10^{-3}	cubic meter, m^3	acre-inch	102.8
35.3	cubic meter, m^3	cubic foot, ft^3	2.83×10^{-2}
6.10×10^4	cubic meter, m^3	cubic inch, in^3	1.64×10^{-5}
2.84×10^{-2}	liter, L (10^{-3} m^3)	bushel, bu	35.24
1.057	liter, L (10^{-3} m^3)	quart (liquid), qt	0.946
3.53×10^{-2}	liter, L (10^{-3} m^3)	cubic foot, ft^3	28.3
0.265	liter, L (10^{-3} m^3)	gallon	3.78
33.78	liter, L (10^{-3} m^3)	ounce (fluid), oz	2.96×10^{-2}
2.11	liter, L (10^{-3} m^3)	pint (fluid), pt	0.473

CONVERSION FACTORS FOR SI AND NON-SI UNITS

To convert Column 1 into Column 2, multiply by	Column 1 SI Unit	Column 2 non-SI Unit	To convert Column 2 into Column 1, multiply by
Mass			
2.20×10^{-3}	gram, g (10^{-3} kg)	pound, lb	454
3.52×10^{-2}	gram, g (10^{-3} kg)	ounce (avdp), oz	28.4
2.205	kilogram, kg	pound, lb	0.454
0.01	kilogram, kg	quintal (metric), q	100
1.10×10^{-3}	kilogram, kg	ton (2000 lb), ton	907
1.102	megagram, Mg (tonne)	ton (U.S.), ton	0.907
1.102	tonne, t	ton (U.S.), ton	0.907
Yield and Rate			
0.893	kilogram per hectare, kg ha^{-1}	pound per acre, lb acre^{-1}	1.12
7.77×10^{-2}	kilogram per cubic meter, kg m^{-3}	pound per bushel, bu^{-1}	12.87
1.49×10^{-2}	kilogram per hectare, kg ha^{-1}	bushel per acre, 60 lb	67.19
1.59×10^{-2}	kilogram per hectare, kg ha^{-1}	bushel per acre, 56 lb	62.71
1.86×10^{-2}	kilogram per hectare, kg ha^{-1}	bushel per acre, 48 lb	53.75
0.107	liter per hectare, L ha^{-1}	gallon per acre	9.35
893	tonnes per hectare, t ha^{-1}	pound per acre, lb acre^{-1}	1.12×10^{-3}
893	megagram per hectare, Mg ha^{-1}	pound per acre, lb acre^{-1}	1.12×10^{-3}
0.446	megagram per hectare, Mg ha^{-1}	ton (2000 lb) per acre, ton acre^{-1}	2.24
2.24	meter per second, m s^{-1}	mile per hour	0.447
Specific Surface			
10	square meter per kilogram, m^2 kg^{-1}	square centimeter per gram, cm^2 g^{-1}	0.1
1000	square meter per kilogram, m^2 kg^{-1}	square millimeter per gram, mm^2 g^{-1}	0.001
Pressure			
9.90	megapascal, MPa (10^6 Pa)	atmosphere	0.101
10	megapascal, MPa (10^6 Pa)	bar	0.1
1.00	megagram per cubic meter, Mg m^{-3}	gram per cubic centimeter, g cm^{-3}	1.00
2.09×10^{-2}	pascal, Pa	pound per square foot, lb ft^{-2}	47.9
1.45×10^{-4}	pascal, Pa	pound per square inch, lb in^{-2}	6.90×10^3

(continued on next page)

Conversion Factors for SI and non-SI Units

To convert Column 1 into Column 2, multiply by	Column 1 SI Unit	Column 2 non-SI Unit	To convert Column 2 into Column 1, multiply by
Temperature			
1.00 (K − 273)	Kelvin, K	Celsius, °C	1.00 (°C + 273)
(9/5 °C) + 32	Celsius, °C	Fahrenheit, °F	5/9 (°F − 32)
Energy, Work, Quantity of Heat			
9.52×10^{-4}	joule, J	British thermal unit, Btu	1.05×10^{3}
0.239	joule, J	calorie, cal	4.19
10^{7}	joule, J	erg	10^{-7}
0.735	joule, J	foot-pound	1.36
2.387×10^{-5}	joule per square meter, J m^{-2}	calorie per square centimeter (langley)	4.19×10^{4}
10^{5}	newton, N	dyne	10^{-5}
1.43×10^{-3}	watt per square meter, W m^{-2}	calorie per square centimeter minute (irradiance), cal cm^{-2} min^{-1}	698
Transpiration and Photosynthesis			
3.60×10^{-2}	milligram per square meter second, mg m^{-2} s^{-1}	gram per square decimeter hour, g dm^{-2} h^{-1}	27.8
5.56×10^{-3}	milligram (H$_2$O) per square meter second, mg m^{-2} s^{-1}	micromole (H$_2$O) per square centimeter second, μmol cm^{-2} s^{-1}	180
10^{-4}	milligram per square meter second, mg m^{-2} s^{-1}	milligram per square centimeter second, mg cm^{-2} s^{-1}	10^{4}
35.97	milligram per square meter second, mg m^{-2} s^{-1}	milligram per square decimeter hour, mg dm^{-2} h^{-1}	2.78×10^{-2}
Plane Angle			
57.3	radian, rad	degrees (angle), °	1.75×10^{-2}

Electrical Conductivity, Electricity, and Magnetism

To convert Column 1 to Column 2, multiply by	Column 1 SI Unit	Column 2 non-SI Unit	To convert Column 2 to Column 1, multiply by
10	siemen per meter, S m^{-1}	millimho per centimeter, mmho cm^{-1}	0.1
10^4	tesla, T	gauss, G	10^{-4}

Water Measurement

9.73 × 10^{-3}	cubic meter, m^3	acre-inches, acre-in	102.8
9.81 × 10^{-3}	cubic meter per hour, m^3 h^{-1}	cubic feet per second, ft^3 s^{-1}	101.9
4.40	cubic meter per hour, m^3 h^{-1}	U.S. gallons per minute, gal min^{-1}	0.227
8.11	hectare-meters, ha-m	acre-feet, acre-ft	0.123
97.28	hectare-meters, ha-m	acre-inches, acre-in	1.03 × 10^{-2}
8.1 × 10^{-2}	hectare-centimeters, ha-cm	acre-feet, acre-ft	12.33

Concentrations

1	centimole per kilogram, cmol kg^{-1} (ion exchange capacity)	milliequivalents per 100 grams, meq 100 g^{-1}	1
0.1	gram per kilogram, g kg^{-1}	percent, %	10
1	milligram per kilogram, mg kg^{-1}	parts per million, ppm	1

Radioactivity

2.7 × 10^{-11}	becquerel, Bq	curie, Ci	3.7 × 10^{10}
2.7 × 10^{-2}	becquerel per kilogram, Bq kg^{-1}	picocurie per gram, pCi g^{-1}	37
100	gray, Gy (absorbed dose)	rad, rd	0.01
100	sievert, Sv (equivalent dose)	rem (roentgen equivalent man)	0.01

Plant Nutrient Conversion

	Elemental	Oxide	
2.29	P	P$_2$O$_5$	0.437
1.20	K	K$_2$O	0.830
1.39	Ca	CaO	0.715
1.66	Mg	MgO	0.602

1 Soil Diversity in the Tropics: Implications for Agricultural Development

H. Eswaran, J. Kimble, and T. Cook

USDA-SCS
Washington, DC

F. H. Beinroth

University of Puerto Rico
Mayaguez, Puerto Rico

The term *laterite* readily comes to mind when the subject of tropical soils is raised. Yet, there is great scope for soil variability in the tropics. The soils that have been referred to as a laterite are, in fact, but of minor extent there. Understanding this diversity and the distribution, characteristics, and processes of the soils of the tropics is obviously of pivotal importance to their wise agricultural development.

Historically, the nature and properties of soils have been an important determinant in cultural and economic development. Many ancient civilizations evolved in the fertile alluvial soils associated with the deltas at the margins of oceans and the floodplains of navigable rivers that were also the avenues for trade, commerce, and communication and thus brought about the emergence of great cities of the past and present. In other areas, such as tropical uplands, the low inherent soil fertility led to farming systems known as *shifting cultivation* or *slash-and-burn agriculture*.

As farmers elsewhere, the farmers of the tropics are keenly aware of soil diversity. This is evidenced by the fact that the "primitive" agriculture of every country is located on the soils with the least constraints. Thus, the more fertile Inceptisols, although they often occur on steep slopes, are preferred to the infertile Oxisols on adjacent plateaus. And the native Indians in the Andean countries preferred volcanic ash soils in the cool uplands to the soils in the hot, humid, and disease-prone Amazon basin.

Since the 1970s, however, population pressures and scarcity of good soils have forced agriculture to expand into hilly, or mountainous, or swamp areas. In parts of East Africa, wildlife is threatened by the encroachment of agriculture into wildlife habitats. In the humid tropics, systematic clearing of the

---

Copyright © 1992 Soil Science Society of America and American Society of Agronomy, 677 S. Segoe Rd., Madison, WI 53711, USA. *Myths and Science of Soils of the Tropics.* SSSA Special Publication no. 29.

forests for agriculture and other purposes has purportedly contributed to the greenhouse effect resulting in global warming. In the semi-arid tropics, uncontrolled irrigation has begun to salinize large areas of land, completely changing the ecosystem.

Two questions arise: (1) Do we know enough about soil diversity? and (2) How do we manage soil diversity so that we have rational land use for competing purposes? These two questions are examined in the context of this publication.

ASSESSING SOIL DIVERSITY

Defining Tropical Soils

Tropical soils may be defined as all those soils that occur in the geographic tropics, that is, in that region of the earth lying between the Tropic of Cancer and the Tropic of Capricorn, also known as the Torrid Zone. The adjective tropical, however, is commonly associated with hot and sultry conditions. Consequently, many think of tropical soils as the soils of the hot and humid tropics only, exemplified by deep, red, and highly weathered soils.

We consider tropical soils to be those that have an "iso" soil temperature regime and lie between the tropics of Cancer and Capricorn. As defined in *Soil Taxonomy* (Soil Survey Staff, 1975), the U.S. system of soil taxonomy, such regimes denote a difference between mean summer and mean winter soil temperatures of 5 °C or less. The implication of an iso-temperature regime, in areas with a mean annual soil temperature >5 °C, is that temperature is not a constraint for most year-round agricultural uses. Iso soil temperatures regimes are almost exclusively confined to the intertropical regions and thus may be considered a soil property that sets the soils of the tropics apart from all other soils. As the tropics comprise approximately 40% of the land surface of the earth, more than one-third of the soils of the world are tropical soils.

Causes of Soil Diversity in the Tropics

Rationalizing soil diversity in terms of the environmental soil-forming factors first postulated by Dokuchaev about 100 yr ago continues to be the unifying philosophy in pedology. In the context of this perspective, the great diversity of soils in the tropics is an inevitable consequence of the enormous diversity of ecosystems found in the intertropical areas. Both the driest and wettest spot on earth are in the tropics, namely in the Atacama Desert, where only sporadic traces of rainfall occur, and on Mt. Waialeala in Hawaii where more than 11 700 mm have been recorded. Mean annual temperatures vary from around 30 °C at the low elevations to below 0 °C on the snowcapped mountains of South America and East Africa. Covariant with this climatic variability, there occur a multitude of ecosystems from deserts to rain forests.

As there was no Pleistocene glaciation in the tropics, many landscapes predate the Quarternary and remnants of peneplains as old as mid-Tertiary, about 20 million yr, are not uncommon. Yet they may occur in juxtaposition with surfaces of recent age. There is also much scope for variation in soil parent material since, with the notable exception of Pleistocene glacial formations such as till and loess, all of the rocks of the temperate zone are also found in the tropics. Moreover, as some soil-forming conditions are unique to the tropics, they have produced soils that can only be found there, (e.g., the Oxisols on ancient geomorphic surfaces).

Some of the oldest geomorphic surfaces are also to be found in the intertropical areas. Examples are the Amazon Shield, the Tertiary surfaces of Africa and southern India. These surfaces have been subject to uplift and tilting with concomitant peneplanation as the most important geomorphic process. As a result, geomorphology is and has been an important control of soil formation in the soils on the older surfaces. Lithological discontinuities marked by stone lines or particle-size differences, are characteristic features of these soils and contribute to the diversity. On the mid- and end-Tertiary surfaces in Africa, the sola and their parent material may have little or no relationship to the underlying rock. In these reworked soils or "sols remaniés," as described in the French literature, the present-day soil is formed on the pedisediment and has little or no relationship to the underlying weathering rock or the rock itself.

In view of the immense environmental diversity encountered in the tropics, often over short distances, the complexity and variability of the resultant soils patterns should come as no surprise. The small island of Puerto Rico may serve as an example: in an area of <9000 km^2, soils representing 10 of the 11 orders currently recognized in *Soil Taxonomy* have been identified.

Cartography and Soil Diversity

Soil diversity is a function of the scale of observation as well as land use. The classes in *Soil Taxonomy* (Soil Survey Staff, 1975) provide for expression of this diversity. Although there is no reliable information on the number of classes occurring in the tropics, particularly at the lower categoric levels, an estimate is given in Table 1-1.

Table 1-1. Estimates of number of soils in each category of the U.S. system of soil taxonomy in the tropics.

Taxonomic level	Estimate of number of soils
Order	11
Suborder	45
Great group	200
Subgroup	1 250
Family	1 000 000
Series	5 000 000
Phases of series	10 000 000

Soil diversity is a function of scale of observation and perhaps a reference level is the minimum decision area (MDA) which is the size of an average small-holder farm in the tropics. For practical purposes this is 1 ha which, when equated to the minimum size of delineation (MSD) on a soil map, requires detailed maps at scales of at least 1:10 000. Variability within MDAs affect farmer performance, and variations between MDAs, the ability to transfer technology and enhance farmer productivity. The number of classes

Fig. 1-1. An example of a detailed soil map of an experimental station (Uganda) (Yost & Eswaran, 1991).

indicate the number of identifiable or mappable entities. Maps at scales of about 1:15 000 (Fig. 1-1) usually depict phases of series but maps at more detailed scales would show greater diversity. Table 1-2 shows the relationship between map scales, minimum size delineation and minimum decision area with respect to scale.

Diversity is also a function of the landscape on which the soils occur. Alluvial terraces and old peneplains tend to be more homogeneous with respect to patterns of soils than steeplands or dissected landscapes. Diversity is also in the eyes of the beholder; some tend to see more variability than others. Soil maps are made for specific objectives and mappers attempt to seek map purity or reduce the variability within map units. Map unit purity and soil variability are obviously objective specific; an area may be considered uniform for rangeland but have considerable variability for cropland.

Many misconceptions regarding the kinds of soils in the tropics arose due to a lack of appreciation of this diversity. This was due to the fact that, until recently, there were only very broad reconnaissance soils maps of the tropical regions. Much of the information about soils in the tropics was accumulated during the period after World War II and came initially from ad hoc observations of a few individuals. The 1:5 000 000 FAO Soil Map of the World aided considerably to erase some of the misconceptions. Though these maps were published in the 1970s, many of the misconceptions have still been carried through. Part of the reason is that apart from soil scientists, many still do not appreciate the terminology introduced in the FAO legend of the Soil Map of the World or in *Soil Taxonomy*. In addition, the amount of information available, as shown by the reliability map for the sheets covering Africa in Fig. 1-2, at the time the FAO map was compiled was inadequate to make a reliable assessment of the soil resources of the continent. Yet even today, this is the best information we have for many parts of Africa. Some countries such as Zambia, Kenya, and Botswana have more recent maps, including maps at larger scales, which differ significantly from the FAO maps. This is, of course, not a criticism of the FAO effort but merely emphasizes the need for more detailed and accurate information

Table 1-2. Map scales, minimum size delineations (MSD) and minimum decision areas (MDA). MSD is the smallest area in which a symbol can be printed on a map. MDA is the smallest area on map from which reliable information can be derived for interpretations. Generally MDA is (4 × MSD).

Map scale	Minimum size delineation		Minimum decision area
	Acres	ha	ha
1:500	0.0025	0.001	0.004
1:2 000	0.040	0.016	0.064
1:5 000	0.25	0.10	0.40
1:10 000	1.00	0.41	1.64
1:20 000	4.00	1.6	6.4
1:100 000	100	40.5	162
1:250 000	623	252	1006
1:1 000 000	10 000	4 000	16 000
1:5 000 000	249 000	101 000	404 000

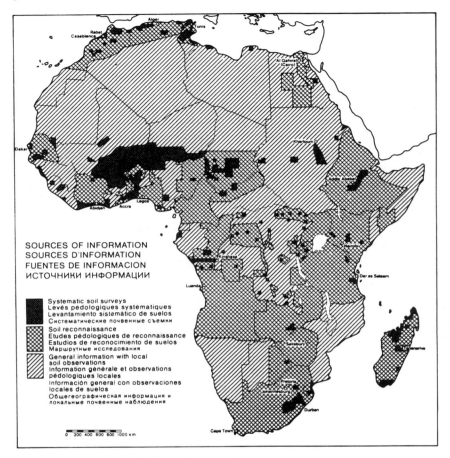

Fig. 1-2. Map showing reliability of FAOs soil map of the world.

that is recognized by all. Nevertheless, it must be realized that basing evaluations by necessity on the generalized world soil map can result in misconceptions and errors.

NOMENCLATURE

Soils nomenclature, particularly as employed by the different kinds of classification systems in existence around the world, has contributed to some of the myths and misconceptions and, in some instances, even confused the issue. Classes in taxonomic systems and map units have observable and measurable properties that fall within class limits that are defined. The same class may show considerable variations in other properties that are not class defining. For some purposes, this variability may be too broad and therefore single property or thematic maps are produced. Good examples are NO_3 status or contents of nutrient elements in soils that are useful information for fertility specialists but are not available from soil survey reports.

Some of the older classification systems, many of which have yet to be modified or improved, employ general and common terms for the classes and differentiate were not strictly defined with respect to class limits. Classification was generally a concensus of opinion (or even the opinion of the leaders) rather than a scientific exercise. This bred a profusion of terms, which though very broad in scope, had good appeal among laypeople. For these reasons, when the U.S. system of soil taxonomy was being developed, a decision was made not to use the old and commonly folkloristic names.

The most notorious of such terms is *lateritic soils* the application of which ranged from any soil of the tropics to the unique feature of a true laterite. Even the term *laterite* had and has a range of meanings and consequently is not employed in the U.S. system of soil taxonomy. However, it still is a popular term among other users of soils, particularly engineers, and thus continues to be a problem today. Names in the U.S. system of soil taxonomy are technical and complicated to the layperson and so many agronomists and plant breeders do not understand them. The situation is worse with economists and sociologists who generally are the decision makers in national or international development programs.

Confusion caused by nomenclature dates back to the 10th century or earlier. The following (cited by Dudal & Eswaran, 1988) illustrates the debate on black soils:

> Democrites state that the best natured soil is the one that takes in rainwater easily, that does not become sticky at the surface, and that does not crack when the rains have ceased. Soils which do not crust as a result of heat are good natured. As a result, says IBN JEDJADJ, to be a good soil it should be neither sticky nor hard. Some have told me, he adds, 'How can the wise DEMOCRITES critisize soils that crack since we see that the soils of the territory of Carmona, which show these features produce higher yields of wheat than those on soils anywhere else?' So, I say that this soil can be depreciated only in comparison with other soils which are of prime quality according to the principles established above. On the other hand, one should not rank the soils that crack among those of first quality just because they produce good wheat. Since a major part of the seeds and plants entrusted to these soils (**VERTISOLS**) do not do well, how could we not give preference to other soils. The black soils with a not too dense texture, which resemble old and well decomposed manure and in which all kinds of seeds and plants succeed (**MOLLISOLS**), should be rated first class on account of their superior quality.
>
> Ibn-Al-Awam
(ca. 10th century)

MYTHS AND REALITY

Myths arise out of misuse of terms, ignorance, and, in the context of developing countries, a desire to achieve the maximum with minimum inputs.

Soil Formation in the Tropics

Knowledge about pedogenesis in the tropics is still incomplete and probably at the relative state where it was 50 yr ago for the temperate areas. This is, of course, related to the difference in the magnitude of pedologic research in the two regions. While the ratio of intertropical to nontropical land areas is about 1:3, that of soil scientists working in the two regions is probably 1:1000.

Pedologists have been accused of looking at the soils of the tropics through "temperate" eyes and studying them in the context of this perspective; the underlying implication being that soil formation in the tropics is somehow very different from soil formation in the higher latitudes. Statements to this effect are only partially correct, however, since the basic processes and reactions of soil formation are the same everywhere. Pedogenetic processes such as lessivage, pedoturbation, eluviation and illuviation, decalcification, and mineralization are universal in nature. Reactions such as hydrolysis, oxidation, and reduction have no geographic boundaries.

Yet, while there is no real difference in the kind of processes operating in the tropics, there may be significant differences in degree. In part of the tropics, the combination of high temperatures, copious amounts of rainfall, and geomorphic stability over millions of years have allowed the pedogenic processes to produce extreme manifestations of soil formation, namely the Oxisols. Figure 1-3, modified from Birkeland (1984) illustrates the differential rates of soil formation in three Orders of soils. The conditions conducive to the formation of such soils are far from ubiquitous, however, and consequently Oxisols account for only about 15% of the intertropical land area. Nevertheless, because Oxisols are unique to this area, the tropics are the only region where all of the 11 soil orders of the U.S. system of soil taxonomy occur.

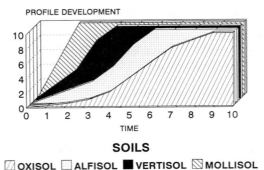

Fig. 1-3. Estimates of time required for formation of different soils. (Modified from Birkeland, 1984.)

Considerations on Soil Formation

Basic differences in soil-forming conditions are related to geology, paleoclimate, geomorphology, and current factors of soil formation. The tropics were free of glaciation and its rejuvenation effects but subject to the pluvial and interpluvial effects. In the tropics, there are soils on old rocks on old landscapes where uplift, erosion, and peneplanation are the main controls of soil formation and these have operated under a high and relatively constant ambient temperature. Soil-geomorphic relationships have a different meaning in the tropics as these must be viewed in the context of paleoclimatic conditions for which there have been few studies. Stone-lines and lithological discontinuities are the norm in the tropics but few studies have focused on these. Many times they are ignored. Horizons of translocated clay accumulation may occur at the top, in the middle, or at the base of the soil and each has a special significance. If an argillic horizon is defined purely on a clay increase, which is the modal situation on the loess of the Midwest, an important genetic property of the tropical soils is lost.

Bioturbation is an important process but on the old landscape of the tropics, it has been in operation for eons. It may have destroyed the evidence of clay translocation (cutans); it has contributed to sorting of particles in the soil; it has modified some features or contributed to the specific morphology of others such as vesicular laterites. As little attention is placed on this aspect of soil formation in the tropics, little is known about these processes.

Recognition and understanding of the processes are important to appreciate the diversity of soils that occur in this environment.

PROPERTIES OF TROPICAL SOILS

Implications for Fertility of Tropical Soils

It is ironic, but agronomists and soil-fertility specialists are often among those who are most ignorant of soil diversity in the tropics. Since the initiation of soil-fertility research decades ago, the same kinds of experiments have been conducted and repeated all over the world with no end in sight. The N-P-K trials keep an agronomist employed, which may be the reason why the experiments never end. In practically every country of the tropics, there is one fertilizer policy for the whole country that may date back to the independence of the country or dictated by a fertilizer supplier. Performance appraisal of the institution is based on the number of soil samples analyzed per year. If all the analyses over the last 50 yr were tabulated, it would not be surprising if essentially every arable soil in the country has been analyzed; yet the process continues.

What many agronomists appear to ignore is that soils are different. Each one or each group of them, requires different management techniques. However, once response patterns have been established and understood, they

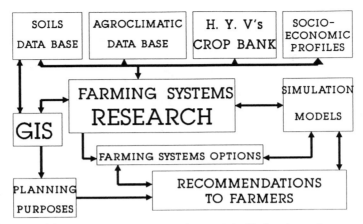

Fig. 1-4. A decision support systems approach to sustainable farming systems research.

can serve to develop models for future management and technology transfer. Figure 1-4 (from Virmani & Eswaran, 1991), shows how databases, particularly soil database, are crucial components in farming systems research and development.

Appreciation of soil diversity in the tropics and aligning management technology to match the specific soil requirements results in considerable savings in foreign exchange, apart from increasing the productivity of agricultural systems.

SOIL DIVERSITY AND SUSTAINABLE AGRICULTURE

What is the soil component in sustainable agriculture? Little is known and no emphasis is given to this important question. International donors such as the U.S. Agency for International Development (AID) are gearing up to meet the challenges of sustainable agriculture but their approach is generally socioeconomic, improving performance, or reducing erosion. They, however, do not seem to appreciate that no one technology will suffice for all situations and that the basic reason is the variability in soil properties. Figure 1-5 (from Eswaran & Virmani, 1990) attempts to illustrate the period a soil would be sustainable under low-input agriculture—the basic tenets of sustainable agriculture. Sustainable agriculture requires not only a knowledge of the resource base but also their attributes and their distribution.

Much of current research activities are confined to experimental stations. Recently, there has been a shift to include farmer's fields. In most developing countries, neither the experimental fields nor the farms are adequately characterized. This raises the question of the utility of such research where the results cannot be transferred to another locality. Apart from this aspect of agronomic research in developing countries, from the point of view of sustainable agriculture, a major shift in research focus is needed. A farm

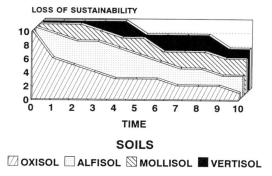

Fig. 1-5. Schematic diagram illustrating period during which low input agriculture is sustainable on different soils.

is a component of a watershed or catchment and from a sustainable agriculture point of view, it is equally important to evaluate other components or segments of the watershed. This calls for a holistic approach or a system-based approach and requires major innovations in research management.

Sustainability also requires evaluations over a time frame that may be of the order of a decade or more. Most donor-funded projects in developing countries do not exceed five and rarely 10 yr. Local staff are few and soon are promoted to other positions. As a result, the life of many field experiments may not be more than 2 or 3 yr. This period is insufficient, particularly in the semiarid tropics, where rainfall can have a coefficient of variation (CV) as high as 30%. There is always the uncertainty if the good responses to the treatments were coincidental, due to the prevailing good weather or if it was a real response to the treatment.

The International Agricultural Research Centers have developed all kind of technologies for their mandated crops. Their general approach is to develop and test a technology under different environments (outreach programs) rather than determining the environments and designing the technologies suited to these situations. This simpler approach is adopted largely because there is insufficient detailed soils information in many of the tropical countries. So should not the donors such as AID be also investing funds to enhance the knowledge of soils of the tropics in their quest for sustainable agriculture?

DEVELOPMENT STRATEGIES

The current knowledge of the soils of the tropics clearly indicates that there is considerable diversity of the soils and the immediate need is to manage this diversity in the context of sustainable agriculture. The approaches are many but if executed in a coordinated manner, the goal is tenable. Some of the prerequisites are discussed below.

Documenting Diversity

As a priority activity, detailed soil surveys must be initiated in all the countries and preferably according to the norms and quality control mechanisms of the Soil Conservation Service of USDA. A trip through the tropical world is an adventure where one can see relicts of soil surveys conducted by fly-by-night consulting companies, expert expatriates many of whom encountered the tropics for the first time.

The U.S. system of soil taxonomy is the only soil classification system designed and improved to meet the conditions of the tropics and which is designed for the explicit purpose of making and interpreting soil surveys. If U.S. funds are involved, it should be mandatory that the U.S. system of soil taxonomy be used and that standards of the SCS of the USDA be adhered to. Many foreign donors are still involved in making soil maps and many decisions, including farm-level decisions, are made on the basis of such maps. There are sufficient general maps of all countries and U.S. development activities should not contribute to more general maps.

Targeting Farming Systems and Conservation Measures

There is increasing awareness of soil degradation today and the concept of sustainable agriculture calls for minimizing degradation. However, the concept of degradation is still subjective and a condition that constitutes degradation on one kind of land may be considered normal elsewhere. In addition, there is still considerable difficulty in differentiating natural degradation from anthropogenic degradation. Some erosion will always occur on steep lands and in the tropics. No catchment can contain all the monsoon rains so some erosion is bound to occur. With no comprehensive base line data for distinguishing the two kinds of degradation in the tropics, decision makers have difficulty in making policy decisions. In fact, Little and Harowitz (1987) report that land use studies of a given area, each purporting to be scientific, can come to widely divergent conclusions about what needs to be done.

A good detailed soil map is the tool for targeting soil conservation measures, recommending farming systems, and for most other uses of soil surveys. Since detailed soil surveys are unavailable in many tropical countries, soil survey interpretation is not appreciated. Instead, land evaluation is in vogue and urged upon by many international organizations, many national policy decisions are being made in the absence of base line data. Land evaluation is a good tool but becomes a meaningless exercise in the absence of reliable and detailed resource information. The fantasy in the developing countries and subscribed to by reputable international organizations and donors, is that land evaluation can be conducted with minimal resource information.

There is a need for a change in the philosophy of farming systems research. Farming systems must recognize soil diversity and must be designed to meet the constraints imposed by this diversity. Matching farming systems to soil properties should become one of the objectives of the research component of sustainable agriculture.

Perceptions of National Decision Makers

In most developing countries, perception of the decision makers is largely influenced by donors and activities of international agricultural organizations. The focus of international research organizations and donors, in the recent past was on improved cultivars—the short cut to increased production. Now with the emphasis on sustainable agriculture, and on productivity rather than on production, there should be a change in some of these priorities and as indicated previously, resource evaluation and conservation need greater thrust. However, this will not happen if donors and the International Agricultural Research Centers (IARCs) do not subscribe to this idea.

Application of Modern Technology

The sparsely populated areas in the less developed countries (LDCs) of the tropics constitute most of the few remaining development frontiers. Growing environmental concerns make it imperative that the inevitable agricultural development of many of these areas in the near future must be guided by rational land use planning. Knowledge about soil is obviously a decisive factor in these considerations and traditionally soil surveys have been made and used to provide the soil information needed for land evaluation. While conventional soil inventories continue to be the backbone of land evaluation, methodologies are now emerging that (i) enhance the quality of soil surveys and the efficiency with which they are made, and (ii) improve their interpretation for agricultural uses.

Some of the state-of-the-art soil survey techniques—the use of video image analysis, ground-penetrating radar, spatial databases—have recently been reported in *Soil Survey Techniques*, SSSA Spec. Publ. 20 (Reybold & Peterson, 1987). Apart from advances in remote sensing, the development of a computer technology known as geographic information systems (GIS) has been especially important. GIS code, store, retrieve, transform, and display maps of land attributes, such as geology, terrain, climate, and vegetation. GIS also have relational database capabilities and thus can manipulate data to create new maps and data elements. Inasmuch as soil variability is predictable to the extent that the factors of soil formation and their influence on pedogenic processes are known, computer programs could be developed that use this information to predict what kind of soil is likely to be found at a specific location.

Computer-aided technologies that can be employed successfully to improve land evaluation include, in addition to GIS, relational databases, crop simulation models, stochastic weather generators, strategy analysis software, pest and pest-loss models, erosion models, farming systems models, and expert systems for various knowledge domains. These techniques can be used to refine yield predictions, evaluate land use alternatives, assess the long-term environmental consequences of agricultural practices, and formulate national and regional production plans. Although at this time, the new technologies are best suited for land evaluation at macroscales where qualitative or semiquantitative assessments suffice, they have nevertheless helped to transform the complex process of land evaluation from an art to a science.

As the degree of sophistication of data management, models, and expert systems increase, so does the amount and specificity of the data required to drive the software. For many areas of the tropics, however, the information is either incomplete or lacking. Burrough (1988) has termed this state of affairs the "parameter crisis"—too many models chasing too few data. There are basically three solutions to this predicament, none of them easy:

1. Collect more data in traditional ways.
2. Make better use of existing data.
3. Generate data with innovative techniques.

The estimation of soil properties, in the absence of site-specific data, appears to be a problem indeed. Yet, assuming that a profile description for the site is available, one could combine certain parameters from this description (horizon depth and thickness, color, texture, and structure) with chemical, physical, and mineralogical properties estimated from characterization data for soils belonging to the same taxon and construct a "synthetic pedon." If the U.S. system of soil taxonomy classification of the soil is not known, an expert system approach can be employed to arrive at an approximation.

But even where soil surveys and characterization data are available, there still is the problem of spatial and temporal soil variability that is particularly important if the new technologies are to be applied at the farm level. In the tropics, it is not unusual that more than one-half of a mapping unit of detailed soils surveys is composed of soils that differ in varying degrees from the typifying soil series. Running a crop model with data derived from the typifying series, may therefore produce quite erroneous results. This dilemma can be solved, however, with a simple expert system. As the range of soils that may conceivably be included in a specific mapping unit of a given area is limited, a pedologist familiar with the area can, by asking a few key questions that a layperson can answer, determine with considerable accuracy what kind of soil occurs at a particular point of a mapping unit.

The emergence of a multitude of agricultural and environmental models that require quantitative soil data has markedly increased the demand for this information in recent years. Soil surveys are the primary source of this information. But as the soil properties routinely determined in soil surveys are not necessarily those required by the models, soil scientists are confronted with the task of developing default procedures to derive the needed parameters from the existing data. Also, it is likely that feedback from the new group of users will impact on the way soil surveys are conducted in the future, particularly with regard to the description and characterization of the tillage zone of the soil.

The application of advanced technologies to deal with the diversity of tropical soils in a development context offers many opportunities but is frequently hampered by the scarcity of site-specific information. The critical need for soil information not only reaffirms the necessity and value of soil survey, but also challenges pedologists to innovatively generate surrogate information to cope with the cases where the pressure for development precludes the generation of resource information with conventional methods.

CONCLUSIONS

As recently as 1972, the prestigious Economic Development Institute of the International Bank for Reconstruction and Development published a treatise that contains the following statement:

> Over a very large part of the humid tropics, the soil has become laterite. That is, through leaching of the main plant foods, the assimilable bases and phosphorus, are removed from the top horizons of the earth. What is left is a reddish mottled clay, consisting almost entirely of hydroxides of iron and alumina whose most distinctive trait is the tendency to solidify on exposure to air... The pure laterites and latosols cover the greater part of the humid tropics and are either agriculturally poor or virtually useless (Karmarck, 1972).

Although clearly a myth, notions of this nature continue to be perpetuated even today. Yet, soils that conform to this description—Oxisols and Ultisols, with plinthite—occupy a very small portion of the tropics, perhaps not more than 2%. In fact and as elaborated above, soil diversity in the geographic tropics is at least as large as that of the temperate zone. This is, of course, patently logical if one considers the enormous variability in the environmental factors that control soil formation in the tropics. Covariant with the taxonomic diversity is a wide range of physical, chemical, and mineralogical properties and resultant soil fertility and production potentials.

No less an authority than Charles E. Kellogg (1967)

> ...fully expects that 'some day' the most productive agriculture of the world will be mostly in the tropics, especially in the humid parts.... Whether 'some day' is 25, 50, 100, or some other number of years from now, depends on how rapidly institutions for education, research, and the other public and private sectors of agriculture will develop.

Today, more than 20 yr after this statement was made, "some day" may have arrived at some plantations and a few other areas. For the most part of the tropics, however, some day remains a futuristic fantasy rather than a realistic scenario. Small-scale farming prevails in most of the region and, as Lal (1987) has pointed out,

> the subsistence farmer who faces famine would consider a successful technology to be the one that produces some yield in the worst year rather than the one that produces a high yield in the best.

But efforts to improve the productivity of subsistence farming are just beginning to get the attention fo the national and international agricultural research centers. Emerging technologies such as agroforestry, alley and multiple cropping, improved genetic material, N_2-fixing trees and crops, and biotechnology hold much promise. In the context of this chapter it should also be mentioned that it would be advantageous if the new farming systems developed are robust enough to be insensitive to the field-scale soil microvariability that tends to exist on tropical farms.

As the world will eventually have to feed 10 billion people, much of the as yet underutilized land in the tropics will have to be brought under cultivation. Can there be sustainable agriculture on these lands? The four basic causes of land degradation—overgrazing on rangeland, overcultivation of cropland, waterlogging and salinization of irrigated land, and deforestation—all result from poor land management and can, therefore, at least in principle, be controlled. The record to date, however, is quite poor. Although effective technologies that prevent or reduce land degradation either exist or are being developed (Postel, 1989), their application is still constrained by institutional and societal barriers. The problem is aggravated by the fact that degradation control crosses all traditional boundaries and that lasting solutions must be rooted as much in social and economic reform as in effective technologies (Postel, 1989). In the tropics, as elsewhere, the prospects for institutionalizing sustainable development strategies are not encouraging.

REFERENCES

Birkeland, P.W. 1984. Soils and geomorphology. Oxford Univ. Press, New York.

Burrough, P.A. 1988. Modelling land qualities in space and time: The role of geographical information systems. p. 91–105. *In* Invited papers presented at the Symp. on Land Qualities in Space and Time, Wangeningen, Netherlands. 22–26 Aug. Agric. Univ. Wageningen, Wageningen, Netherlands.

Dudal, R., and H. Eswaran. 1988. Distribution, properties and classification of Vertisols. p. 1–22. *In* L.P. Wilding, and R. Puentes (ed.) Vertisols: Their distribution, properties, classification and management. Publ. Soil Manage. Support Serv., Washington DC.

Kamarck, A.M. 1972. Climate and economic development. EDI Seminar Paper no. 2. Economic Development Inst., Int. Bank for Reconstruction and Development, Washington, DC.

Kellogg, C.E. 1967. Comment. p. 232–233. *In* H.M. Southworth and B.F. Johnston (ed.) Agricultural development and economic growth. Cornell Univ. Press, Ithaca, NY.

Lal, R. 1987. Managing the soils of sub-Saharian Africa. Science 236:1069–1076.

Little, P.D., and M.H. Harowitz. 1987. Social science perspectives on land ecology and development. p. 1–3. *In* P.D. Little et al. (ed.) Lands at risk in the third world: Local level perspectives. Westview Press, Boulder, CO.

Postel, S. 1989. Halting land degradation. p. 21–40. *In* L.R. Brown (ed.) State of the world 1989. A Worldwatch Institute report on progress toward a sustainable society. W.W. Norton and Co., New York.

Reybold, W.U., and G.W. Peterson (ed.). 1987. Soil survey techniques. SSSA Spec. Publ. 20. SSSA, Madison, WI.

Soil Survey Staff. 1975. Soil taxonomy: A basic system of soil classification for making and interpreting soil surveys. USDA-SCS Handb. 436. U.S. Gov. Print. Office, Washington, DC.

Virmani, S.M., and H. Eswaran. 1991. Concepts for sustainability of improved farming systems in the semi-arid regions of developing countries. Int. Conf. "Soil Quality in semi-arid agriculture," Saskatoon, Canada. (In press.)

Yost, D., and H. Eswaran. 1991. Major land resource areas of Uganda. (In press.)

2 Organic Matter Dynamics in Soils of the Tropics—From Myth to Complex Reality

D. J. Greenland

CAB International
Wallingford, Oxford, England

A. Wild and D. Adams

University of Reading
Reading, England

Myths have always clung, and still cling, to the importance of organic matter in soils. Although its significance is sometimes exaggerated, there is a well-established scientific basis for its importance. It is not necessary to reiterate the significant functions of soil organic matter in this chapter. It is, however, essential to state that in many tropical soils with low-activity clay fractions its influence on crop production is particularly high. Many early reports about tropical soils emphasized their red colors and apparent low organic matter status, and the difficulty of maintaining adequate organic matter levels. The problems of sustaining productivity under arable farming conditions led to the impression, and the myth, that the "humus" of tropical soils was different and of poorer quality from that of soils of the temperate zone. The higher rate of turnover of organic matter under higher temperature conditions, and hence the greater problem of maintaining organic matter levels is not a myth. The higher rate of turnover has, however, led to the myths, both about quantity and quality of organic matter in soils of the tropics. There is now sufficient evidence that there is no difference in quality and effectiveness of humus in tropical and temperate soils. A desirable optimum level for organic matter content exists for most soils. Hence, an understanding of organic matter dynamics is a subject of considerable and continuing interest.

THE MYTHS AND THEIR SOURCES

The Quantity Myth

This may be stated "There is negligible organic matter in tropical soils, and the rate of oxidation at the temperatures prevailing is so high that there

---

Copyright © 1992 Soil Science Society of America and American Society of Agronomy, 677 S. Segoe Rd., Madison, WI 53711, USA. *Myths and Science of Soils of the Tropics.* SSSA Special Publication no. 29.

is no point in attempting to increase it." Origins of the myth lie in explorers' tales of the red color showing little indication of the black or brown colors of the better soils of the temperate zones, and early attempts at cultivating these soils, when they were found to lack mellowness, or be extremely harsh when compared with temperate zone soils—good tilth and soft mellowness being properties associated with relatively high levels of organic matter. These simple and straightforward observations, which are correct for the soils on which they were made, were perpetuated by mistaken reports such as that of Lundegardh (1931). He stated that "In the tropics...decomposition proceeds...more rapidly than the formation of new organic material," which creates a mystery as to how it occurs at all. Presumably, he was referring only to nonhumified material.

Support for the myth has come subsequently from various reports. For example, Beirnaert (1941) who, by comparing two adjacent soil areas, one under forest, the other cleared and kept bare, concluded that 6 t of humus and old roots were oxidized in 1 mo, and that the annual rate of loss of humus was 12% of the total. This rate is much in excess of what has been found in more rigorous studies of organic matter loss made on the same plot of soil over successive years.

The Quality Myth

This may be stated: "Humus as we know it is not formed in the tropics—what is there, if any, does little good." This myth seems to have arisen from agriculturalists' experience of the difficulties in managing red soils in the tropics. The intractability of the very black soils—the black cotton soils which contrasted so sharply with similarly colored chernozems and rendzinas may also have contributed. Their color suggested they should be similar in organic matter content and as easy to manage and as productive as the chernozems. They are, in fact, exceptionally difficult to manage and low in organic matter content in spite of their color.

The myth arose at a time when soil scientists were avidly seeking to prove that active humus could be extracted from the soil with a particular reagent, the nonextracted material being the inert, putatively valueless organic material. One extractant supposed able to isolate such active humus was ammonium oxalate, the use of which was strongly urged by Chaminade, amongst others. But Fauck (1956) noted that after analyzing several thousands of samples from West Africa, "no significant correlation could be obtained with yields, nor with carbon, whether after green manures, after fallow, after groundnuts, after cereals or under forest." Bremner (1954) concluded in 1954 that "non-isolative methods are of very uncertain and limited value" and urged the search for identifiable chemical compounds. This has also not proved a particularly profitable route to follow.

Deflating the Myths

The myths have never been exploded, but have been deflated by the continuing growth of factual information. Vageler (1930) was already endeavoring to deflate the myths. He states, "The lack of humus coloration not only in the subsoil but also in the surface layers of the profile has led to the view that tropical soils are notably low in humus, and that there is no need to bother about humus and no danger in destroying humus by intensive cultivation and clean weeding. Seldom have appearances led to a more mistaken conclusion. In the first place, the humus content of soil in tropical climates, especially in regions of heavy rainfall, is by no means low, and in the second place, humus is just as important a factor in conditioning tropical land as in temperate climates." Nevertheless, arguments about continuous cultivation of soils in the tropics and the importance of organic matter to such continuous cultivation have continued.

Although the myths have not been exploded, there has been an explosion of knowledge about the soils of the tropics. The myths grew up around experience with a restricted range of soils, and in a limited range of environments. Although there has been a continuing tendency to extrapolate results obtained on a single soil type in one environment to all tropical soils, differences in behavior are becoming more widely understood.

Jenny's classical work (Jenny, 1930) on the relation between mean annual temperature and organic matter content, established for a transect from north to south in the USA, provided a considerable degree of respectability to the idea that "tropical soils" would have low organic matter contents. His subsequent studies, however, showed that the relationship established in the USA could not be extrapolated to tropical areas (Jenny et al., 1948; Jenny, 1949, 1950, 1961) as the organic matter contents of tropical soils were generally higher than predicted. The comparison should of course be made between soils of the same Order, and of comparable history of use. Sanchez (1976) has made the comparison for soils of the same order (Table 2-1) and found levels of organic matter to be comparable between soils in the tropical and temperate zones. Tanaka et al. (1984, 1986) also comment on the similarity of organic matter levels between soils of the Amazon Basin and soils of the same Order in Japan.

Table 2-1. Comparison of average organic C contents of several soil orders in the USA, Brazil, and Zaire. Each figure is the average of 16 randomly chosen profiles (%) (Sanchez, 1976).

Soil order	USA	Brazil	Zaire	Means
0–15 cm depth				
Mollisols	2.44	--	--	2.44
Oxisols	--	2.01	2.13	2.07
Ultisols	1.58	1.61	0.98	1.39
Alfisols	1.55	1.06	1.30	1.30
	LSD (0.05) = 0.38			
0–100 cm depth				
Mollisols	1.11	--	--	1.11
Oxisols	--	1.07	1.03	1.05
Ultisols	0.49	0.88	0.45	0.61
Alfisols	0.52	0.53	0.55	0.53
	LSD (0.05) = 0.19			

Source: Wade and P. Sanchez (unpublished data).

Van't Hoff's law—the rate of any chemical reaction approximately doubles for every 10 K rise in temperature—cannot be denied, even for microbiologically induced changes. The law is subject to the restriction that all other factors that can influence the rate of reaction must be held constant. Amongst

Table 2-2. Estimates of litter fall in tropical forests (t ha^{-1} yr^{-1}). Leaf litter is presented separately from total litter which also includes fruit and flower parts and twigs (UNESCO, 1978).

Formation type	Location	Leaf litter	Total litter
Evergreen	Banco, Ivory Coast	8.2	11.9
Evergreen	Yapo, Ivory Coast	7.1	9.6
Evergreen	Olokemeji, Nigeria	--	7.2
Moist semi-deciduous	Omo, Nigeria	--	4.6
Secondary moist semi-deciduous	Nigeria	--	5.6
Secondary forest (40 yr)	Kade, Ghana	6.9	10.5
Semi-deciduous	Tafo, Ghana	--	20.9
Mixed forest	Yangambi, Zaire	--	12.4
Brachystegia	Yangambi, Zaire	--	12.3
Macrolobium	Yangambi, Zaire	--	15.3
Musanga	Yangambi, Zaire	--	14.9
Dry evergreen	Lubumbashi, Zaire	4.7	9.2
Miombo forest	Lubumbashi, Zaire	3.0	4.0
Miombo	Lubumbashi, Zaire	2.9	7.5
Riparian	Lubumbashi, Zaire	4.5	5.9
Tropical moist	Darien, Panamá	--	11.3
Premature wet	Darien, Panamá	--	10.5
Gallery	Darien, Panamá	--	11.6
Second growth	Canal Zone, Panamá	--	6.0
Rain forest	Colombia	--	8.5
Rain forest	Colombia	--	10.1
Amazonian upland	Manaus, Brazil	5.6	11.3
Evergreen seasonal (*Mora*)	Trinidad	7.0	--
Lower montane	El Verde, Puerto Rico	4.8	11.4
Montane	Rancho Grande, Venezuela	--	7.8
Dry forest	Calabozo, Venezuela	--	8.2
Lowland dipterocarp	Malaysia	--	7.2
Lowland dipterocarp	Malaysia	--	5.5
Upland dipterocarp	Malaysia	--	6.3
Secondary forest	Malaysia	--	8.3
Secondary forest	Malaysia	--	10.5
Secondary forest	Malaysia	--	14.4
Lowland dipterocarp	Pasoh, Malaysia	8.3	12.6
Evergreen-gallery	Thailand	--	25.3
Temperate evergreen	Thailand	--	18.9
Dipterocarp savanna	Thailand	--	7.8
Mixed savanna	Thailand	--	8.0
Equatorial rain	Khao Chong, Thailand	11.9	23.2
Terminalia-Shorea	Chakia, India	6.2	--
Tectona	Chakia, India	5.0	--
Diospyros-Anogeissus	Chakia, India	4.2	--
Shorea-Buchanania	Chakia, India	3.1	--
Butea	Chakia, India	1.0	--
Dry deciduous	Varanasi, India	--	7.7
Montane (2100 m)	Gundar, India	3.9	5.5
Chir plantation	Dehra Dun, India	--	7.8
Teak plantation	Dehra Dun, India	--	7.8
Sal plantation	Dehra Dun, India	--	10.9

such factors affecting C oxidation rates in soils are: clay type and content, pH, and moisture conditions. The fact that soils belong to the same order, or even suborder, family or series, does not mean that pH and clay type and content in the surface horizon are similar. Also, because the soil organic matter level reflects the balance of losses and additions, even when other factors influencing the rate of decomposition are equal, it is not possible to observe the effect of temperature on losses of C unless the rate of addition of organic material to the soils being compared is also considered. Rates of addition of organic matter to soils under tropical rain forest are exceptionally high. UNESCO (1978) gives an average litter addition of 10.4 t ha^{-1} (Table 2-2). More recently, Proctor (1984) gave 8.3 t ha^{-1} based on more than 100 determinations. This is rather less than Vageler's original estimate of more than 100 t ha^{-1} fresh weight, but still much higher than in temperate forests (Rennie, 1955). Thus, the level of organic matter in soils under tropical forest may be expected to be high rather than low, although changes in land use, and particularly cultivation that enhances the rate of decomposition, may be expected to reduce the level of organic matter rapidly.

Amongst other observations on soils in the tropics that helped to deflate the quantity myth were the high levels of organic matter in Andepts and other soils containing a high proportion of allophane in the clay fraction (Bornemisza & Pineda, 1969) and often moderately high levels in certain Oxisols (Sanchez, 1976) and Rhodustalfs (Moormann, 1981). Oxisols and Rhodustalfs derived from basic or ultrabasic parent materials have consistently high organic matter contents. Like the Andepts, they contain significant amounts of noncrystalline material in the clay fraction. Sombroek and Siderius (1981) have rightly urged a reconsideration of the classification of these soils.

REPLACING THE MYTHS WITH REALITY

The Realities of Quantities of Organic Matter

The enormous accumulation of analytical data about tropical soils now shows clearly that the quantities of organic matter in tropical soils cover a wide range. The amount of organic matter at any one time is the result of the rates of addition and loss. Figure 2-1 illustrates this. As noted above, under natural forest vegetation the rate of input will be high, so that the equilibrium level is high in spite of the fact that the decomposition rate is also rather high. In cultivated soils, however, the rate of organic matter return is much lower. Organic matter will then fall to lower equilibrium levels than in comparable arable soils in the temperate zone. Considerable data exist to show the decline in organic matter content of cultivated soils in the tropics. Some of the data are summarized in Fig. 2-2.

It is well established that fresh organic material breaks down rapidly after addition to the soil, and much more so than humified material. As was recognized by those who created the quality myth, this is important to soil productivity because much of the supply of N and S to crops depends on

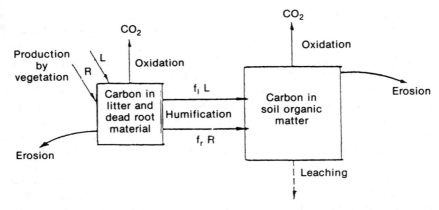

Fig. 2-1. Processes in the addition and loss of carbon from soils (Greenland & Nye, 1959).

their mineralization during the breakdown of organic matter. The composition of the organic matter added to the soil also influences its rate of breakdown. Litter and root material, for instance, decompose rather differently.

As the C/N ratio is relatively constant in soil organic matter, N changes in the soil follow C changes. Bartholomew (1975) and Greenland (1985) have summarized much data on the dynamics of soil N in tropical soils. The changes in N are less distorted by the influence of fresh organic material in the soil, as the proportion of N in the soil in such material is lower than that of C.

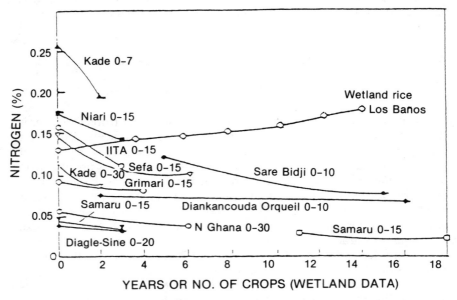

Fig. 2-2. Nitrogen changes in upland soils after clearing from natural vegetation (Nye & Greenland, 1960; Jones & Wild, 1975; Juo & Lal, 1977), and of wetland soil cultivated to rice (Greenland & Watanabe, 1982). (From Greenland, 1985).

The external factors that influence the rate of mineralization of soil organic matter include temperature and rainfall, the vegetation in so far as it modifies soil temperature and biological activity in the soil, and the composition of the organic material entering the soil and soil disturbance by cultivation. The intrinsic properties include clay content, clay type, drainage, acidity, and nutritional status.

Several authors have described the relation between clay content and organic matter level. For instance, for soils of the savanna region of West Africa, Jones (1973) found the relation

$$\%C = 0.341 + 0.0273\ Y$$

where $Y = \%$ clay. There is certainly an effect of clay type as well as clay amount on the retardation of organic matter decomposition in soils. It is probable that this is partly an extension of the effect of amount, due to the fact that 2:1 clay minerals generally have greater surface areas than 1:1 minerals. The protective effect may be due to adsorption, but may also be associated with the higher proportion of pores $< 1\ \mu m$ in equivalent diameter, in which humic materials are inaccessible to microbial attack. The highest surface areas and proportion of very fine pores occur in soils containing amorphous inorganic materials. Tropical Andepts are, of course, renowned for their exceptionally high organic matter contents (Quantin, 1972; Leamy et al., 1980; Martin & Haider, 1986; Boudot et al., 1986).

It has sometimes been suggested that protection arises from interlamellar adsorption by 2:1 expanding lattice clays (Allison, 1973). The evidence of interlamellar adsorption of soil organic materials in this way is meagre. In fact, Vertisols in tropical conditions have low contents of organic matter ($\%C < 1.0$) in spite of high contents of clay, most of which is smectitic. Thus, there must be a further factor determining the stability of organic matter in Andepts. Wada (1980) and Virakornphanich et al. (1988) present evidence that this is due to association with active aluminum and iron hydrous oxides. This mechanism might also partly account for the relatively high stability of organic matter in some Oxisols and other soils whose clay fraction is dominated by hydrous oxides, and in very acid soils. The effect of strong acidity was first reported by Hardon (1936) and recent data bear out his earlier findings (Fig. 2-3). The full complexity of the clay-organic complex and the reasons for the stability of organic matter in the complex have still to be fully elucidated (Tate & Theng, 1980; Goh, 1980; Oades, 1984).

Drainage usually implies poorer soil aeration, and a slower rate of organic matter decomposition. In temperate soils, this leads almost invariably to the accumulation of organic material in soils that are poorly drained. In the tropics, this is not equally true. Mohr (1922) noted that at above 30 °C the rate of decomposition of organic matter by anaerobic organisms became sufficiently rapid so that poor drainage did not necessarily lead to accumulation of organic matter (Fig. 2-4). Neue and Scharpenseel (1987) have recently confirmed the high rates of decomposition that can be attained in saturated soils in tropical conditions, by following the decomposition of

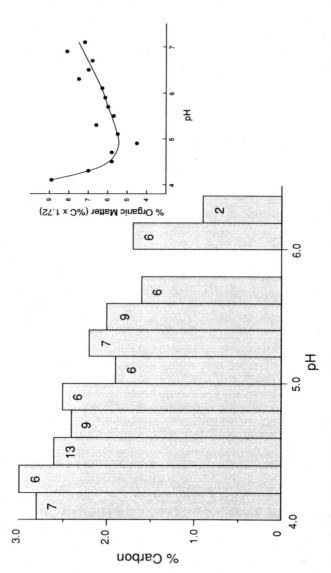

Fig. 2-3. (*Left*) Relation between pH and organic C content for A1 horizons of 77 lowland acid soils from West Africa (S Nigeria and Sierra Leone) and India. Soils are grouped in ranges 4.0 to 4.2, 4.2 to 4.4, etc. Numbers represent numbers of soils in each group. (Data from Odell et al., 1974; Indian Soc. Soil Sci., 1976; and Greenland, 1981.)

(*Right*) Relation between pH and organic matter content for 971 soil samples (0–4 in.) from southern Sumatra (Hardon, 1936).

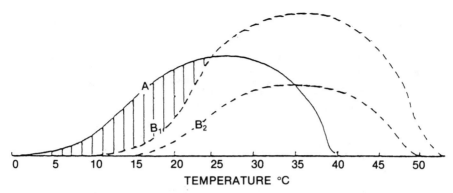

Fig. 2-4. Essential features of Mohr's Diagram relating to organic matter. Curves represent—A, synthesis by plants; B1, destruction by aerobic bacteria; B2, destruction by anaerobic bacteria. The shaded portion indicates conditions favorable for "accumulation".

Note: The curve A represents units that are totally different from those of B1 and B2. The fact that A and B1 happen to intersect at 25 degrees rather than 15 or 35 has, therefore, no significance. The diagram does, however, illustrate the fact that organic matter accumulates especially at low temperatures and in anaerobic conditions, but that above about 30°C the accumulation is likely to be small even in conditions where anaerobic bacteria are dominant.

(From Leeper, 1938, after Mohr, 1922).

^{14}C-labelled rice (*Oryza sativa* L.) straw under continuously flooded conditions for 4 yr (Fig. 2-5). The low organic matter contents of many wetland areas of tropical Asia where paddy rice is grown (Kyuma, 1985) are of course further testimony to the relatively high rates of decomposition. As Neue and Scharpenseel (1984) have said, much further information is needed on the mineralization of organic material in tropical soils under poorly drained conditions.

The Realities of Quality of Organic Matter

The search for something that can be identified as "true humus" was long and arduous, and should have been abandoned sooner than it was. "Humus" as an identifiable substance that can be separated from other constituents of soil organic matter is a myth that has confounded soil scientists in the temperate as well as tropical zone for too many years. There are, however, parts of soil organic matter that mineralize at different rates. In so far as the release of N and S in plant assimilable form reflects the quality of humus, the most readily mineralized part of the organic matter is the most valuable. As the mineralization process is also that from which the soil population derives its energy, the mineralizable part of organic matter also reflects the level of biological activity in the soil. Soil organic matter, however, is also important to soil productivity in other ways. In red soils of the tropics, it is important to physical properties, phosphate sorption, micronutrient availability, and often critically important to cation retention (Greenland, 1986).

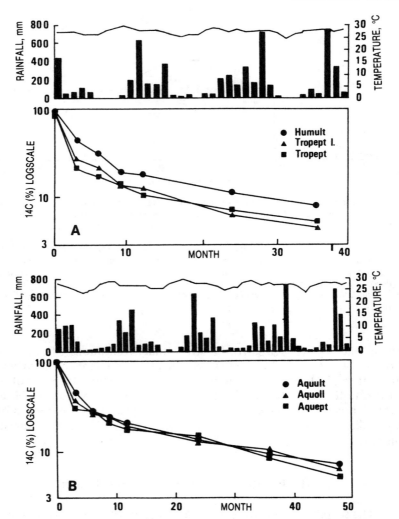

Fig. 2-5. Decomposition pattern of ^{14}C-labelled rice straw: (A) in tropical upland soils, (B) in tropical lowland soils (Neue & Scharpenseel, 1987).

Attempts were partially successful to relate the readily mineralized part of the organic matter to freshly added organic material, separable as the light fraction not complexed with heavier inorganic material. Greenland and Ford (1964) showed that the material separable on the basis of specific gravity <2 was a complex mixture of humified and nonhumified material. Ford and Greenland (1968) concluded that while the light fraction was a more labile fraction of the organic matter than more humified material, both fractions are involved in mineralization, their relative importance varying with soil type. Similarly, while the polysaccharide component of soil organic matter plays an important role in interparticle bonding and aggregate stabilization, it does not play the sole role. Fungal hyphae and other material are also important (Tisdall & Oades, 1982).

DYNAMICS OF SOIL ORGANIC MATTER TURNOVER

The equation:

$$I = A - kC$$

where
- I = annual increase in soil C,
- A = annual addition,
- C = total C content, and
- k = a decomposition constant

is useful as a simple description of the rate of C loss from soils that have been in cultivation for a few years, and which receive relatively small amounts of organic material. However, in reality, both A and C are composed of many fractions, which decompose at different rates, and a more complex model is needed to describe successfully the turnover of soil C. Several such models have been developed. Most of them divide the soil organic matter (SOM) into several compartments, with each compartment having a characteristic decomposition rate. These rates can then be modified by including climatic data and information on soil properties that affect the decomposition rates.

Earlier models of this type (Hunt, 1977; Jenkinson & Rayner, 1977) contain ideas and equations used in the development of more recent models. Examples of these are the Century model, (Parton et al., 1983), Van Veen and Paul's (1981) long-term organic C turnover model (validated by Voroney et al., 1981), and the model of Jenkinson et al. (1987) which is based on that of Jenkinson and Rayner (1977). A diagram of the flow of C through the updated model may be seen in Fig. 2-6.

In the 1987 model, organic matter enters the system via either the decomposable (DPM) or resistant plant material (RPM) compartments, passing through these only once. The contents of these compartments and the soil biomass (divided into zymogenous [BIOZ] and autochthonous [BIOA] compartments) decompose by process A to form CO_2, BIOZ, and humus (HUM). HUM decays more slowly by process B to form CO_2, BIOA, and HUM. In addition, the soil is assumed to contain a small organic fraction (IOM), which is inert to biological attack.

The model of Jenkinson et al. (1987) takes into account climatic data such as temperature and soil moisture, and some soil characteristics including the clay content but not clay type, nor acidity. Also included is a plant retainment factor given the value 0.6 if there are plants growing in the soil, and 1.0 if the soil is kept clear of vegetation, since decomposition proceeds more quickly in cultivated bare soils.

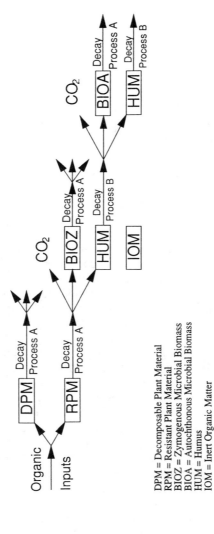

Fig. 2-6. Carbon flow in soils (Jenkinson et al., 1987). DPM = decomposable plant material, RPM = resistant plant material, BIOZ = zymogenous microbial biomass, BIOA = autochthonous microbial biomass, HUM = humus, and IOM = inert organic matter.

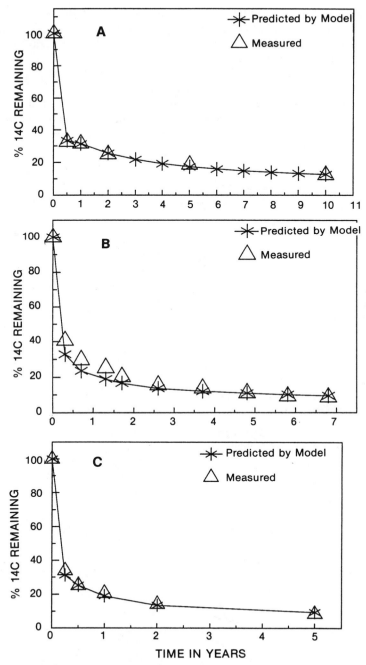

Fig. 2-7. Decomposition of ^{14}C-labelled plant material at (A) Rothamsted Experimental Station (Jenkinson, 1977), (B) Toboga, Costa Rica (Gonzalez & Sauerbeck, 1982), and (C) Ibadan, Nigeria (Jenkinson & Ayanaba, 1977). Diagrams from Jenkinson et al. (1991).

This model has now been updated (Jenkinson, 1990) and is currently being tested against measured data from several sites around the world to see how well it simulates the turnover of soil organic matter in different conditions, in both temperate and tropical zones (Jenkinson et al., 1991). Some of the results may be seen in Fig. 2-7 that shows how the more rapid decomposition in some soils of the tropics can be simulated.

Changes in soil C (SOM) level arising from changes in input of organic matter are illustrated in Fig. 2-8 using constants appropriate for tropical con-

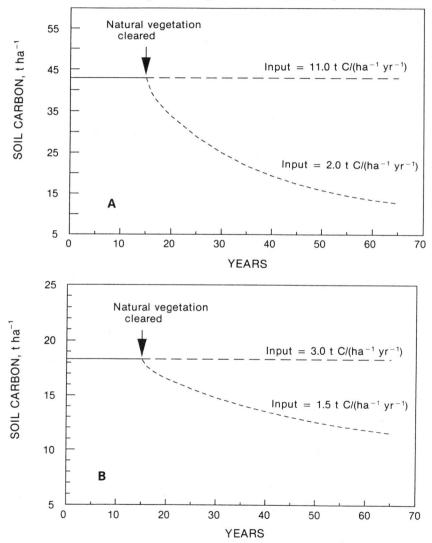

Fig. 2-8. Simulation of the effect of clearance and cultivation on contents of tropical soils using the updated Jenkinson model (A) Wet tropics—Forest, and (B) Seasonally arid—Savanna (Jenkinson, 1990).

ditions. Soil organic C is predicted to fall from 43 to 25 t ha^{-1} in 15 yr, with continuing inputs of 2.0 t ha^{-1} instead of 11.0 t ha^{-1} of C per year. These levels correspond approximately to humid tropical conditions changing from forest cover to arable crop production. For drier, savanna conditions a fall from 18 to 14 t ha^{-1} in 15 yr is predicted. These simulated data may be compared with the real data shown in Fig. 2-2.

CONCLUSION

There are more than sufficient facts to dispel the myths about organic matter and its dynamics in soils of the tropics. Generalizations about tropical soils are unlikely to have wide applicability, because of the diversity of soils and the factors affecting organic matter dynamics. We do have models able to take account of many of the factors influencing organic matter dynamics, but more data are needed to verify the models for different soil conditions, and establish the values of the constants to be included for different soils and climates.

REFERENCES

Allison, F.E. 1973. Soil organic matter and its role in crop production. Elsevier, Amsterdam.

Bartholomew, W.V. 1975. Soil nitrogen changes in farming systems in the humid tropics. p. 27-44. *In* A. Ayanaba and P.J. Dart (ed.) Biological nitrogen fixation in farming systems of the tropics. Wileys, Chichester, UK.

Beirnaert, A. 1941. La technique culturale sous l'equateur. Influence de la culture sur les reserves en humus et en azote des terres equatoriales. Inst. Natl. pour les Etudes Agron. Congobelge (INEAC) Ser. Tech. No. 26. INEAC, Brussels.

Bornemisza, E., and R. Pineda. 1969. The amorphous minerals and the mineralisation of nitrogen in volcanic ash soils. p. B 7.1-7.7. *In* Report of panel on soils derived from volcanic ash in Latin America. Inter-American Inst. of Agric. Sci., Turrialba, Costa Rica.

Boudot, J.P., B.A. Bel Hadj, T. Chone, and B.A.B. Hadj. 1986. Carbon mineralisation in Andosols and aluminum-rich highland soils. Soil Biol. Biochem. 18:457-461.

Bremner, J.M. 1954. A review of recent work on soil organic matter. J. Soil Sci. 5:214-232.

Fauck, R. 1956. Evolution of soils under mechanized cultivation in tropical areas. Trans. 6th Int. Congr. Soil Sci. E:593-596.

Ford, G.W., and D.J. Greenland. 1968. The dynamics of partly humified organic matter in some arable soils. Trans. 9th Int. Congr. Soil Sci. 2:403-410.

Goh, K.M. 1980. Dynamics and stability of organic matter. p. 373-396. *In* B.K.G. Theng (ed.) Soils with variable charge. New Zealand Soc. Soil Sci., Lower Hutt, NZ.

Gonzalez, A.M.A., and D.R. Sauerbeck. 1982. Decomposition of ^{14}C labelled plant residues in different soils and climates of Costa Rica. p. 141-146. *In* C.C. Ceri et al. (ed.) Proc. Regional Colloquium on Soil Organic Matter Studies. CENA/USP; PROMOCET, São Paulo, Brazil.

Greenland, D.J. (ed.). 1981. Characterisation of soils in relation to their classification and management for crop production: Examples from some areas of the humid tropics. Oxford Univ. Press, Oxford, UK.

Greenland, D.J. 1985. Nitrogen and food production in the tropics: Contributions from fertilizer nitrogen and biological fixation. p. 9-38. *In* B.T. Kang and J. van der Heide (ed.) Nitrogen management in farming systems in humid and subhumid tropics. Inst. for Soil Fertility, Haren, Netherlands.

Greenland, D.J. 1986. Effects of organic matter on the properties of some red soils. p. 262-273. *In* Proc. Int. Symp. on Red Soils. Science Press, Beijing, China and Elsevier, Amsterdam.

Greenland, D.J., and G.W. Ford. 1964. Separation of partially humified organic materials from soils by ultrasonic dispersion. Trans. 8th Int. Congr. Soil Sci. Bucharest 3:137–148.

Greenland, D.J., and P.H. Nye. 1959. Increases in the carbon and nitrogen contents of tropical soils under natural fallows. J. Soil Sci. 10:284–299.

Greenland, D.J., and I. Watanabe. 1982. The continuing nitrogen enigma. Trans. 12th Int. Congr. Soil Sci., 5:123–137.

Hardon, H.J. 1936. Faktoren die het organische stof-en het stikstofgehalte van tropische gronden beheersen. Korte Meded. Alg. Proefsta. Landb.

Hunt, H.W. 1977. A simulation model for decomposition in grasslands. Ecology 58:469–484.

Indian Society of Soil Science. 1976. Acid soils of India, their genesis, characteristics and management. Indian Soc. Soil Sci., New Delhi.

Jenkinson, D.S. 1977. Studies on the decomposition of plant material in soil. V. The effects of plant cover and soil type on loss of carbon from ^{14}C labelled ryegrass decomposing under field conditions. J. Soil Sci. 28:424–434.

Jenkinson, D.S. 1990. Turnover of organic carbon and nitrogen in soil. Phil. Trans. R. Soc. London Ser. B 329:361–368.

Jenkinson, D.S., D.E. Adams, and A. Wild. 1991. Model estimates of CO_2 emissions from soil in response to global warming. Nature (London) 351:304–306.

Jenkinson, D.S., and A. Ayanaba. 1977. Decomposition of ^{14}C labelled plant material under tropical conditions. Soil Sci. Soc. Am. J. 41:912–915.

Jenkinson, D.S., P.B.S. Hart, J.H. Rayner, and L.C. Parry. 1987. Modelling the turnover of organic matter in long-term experiments at Rothamsted. INTECOL Bull. 15:1–8.

Jenkinson, D.S., and J.H. Rayner. 1977. The turnover of soil organic matter in some of the Rothamsted classical experiments. Soil Sci. 123:298–305.

Jenny, H. 1930. A study of the influence of climate upon the nitrogen and organic matter content of the soil. Missouri Agric. Exp. Stn. Res. Bull. 152.

Jenny, H. 1949. Comparative study of decomposition rates of organic matter in temperate and tropical regions. Soil Sci. 68:419–432.

Jenny, H. 1950. Causes of the high nitrogen and organic matter content of certain tropical forest soils. Soil Sci. 69:63–69.

Jenny, H. 1961. Comparison of soil nitrogen and carbon in tropical and temperate regions as observed in India and America. Missouri Agric. Exp. Stn. Bull 765.

Jenny, H., F. Bingham, and B. Padilla-Saravia. 1948. Nitrogen and organic matter contents of equatorial soils of Colombia, South America. Soil Sci. 66:173–186.

Jones, M.J. 1973. The organic matter content of the savanna soils of West Africa. J. Soil Sci. 24:42–53.

Jones, M.J., and A. Wild. 1975. Soils of the West African Savanna. Tech. Comm. 55. Commonwealth Bureau of Soils, Harpenden, UK.

Juo, A.S.R., and R. Lal. 1977. The effect of fallow and continuous cultivation on chemical and physical properties of an Alfisol in Western Nigeria. Plant Soil 47:574–584.

Kyuma, K. 1985. Fundamental characteristics of wetland soils. p. 191–206. *In* Wetland soils: Characterisation, classification, and utilisation. Int. Rice Res. Inst., Los Baños, Philippines.

Leamy, M.L., G.D. Smith, F. Colmet-Daage, and M. Otowa. 1980. The morphological characteristics of Andisols. p. 17–34. *In* B.K.G. Theng (ed.) Soils with variable charge. New Zealand Soc. of Soil Sci., Lower Hutt, NZ.

Leeper, G.W. 1938. Organic matter of soils as determined by climate. J. Australian Inst. Agric. Sci. 4:145–147.

Lundegardh, H. 1931. Environment and plant development. Arnold, London.

Martin, J.P., and K. Haider. 1986. Influence of mineral colloids on turnover rates of soil organic carbon. p. 283–304. *In* P.M. Huang and M. Schnitzer (ed.) Interactions of soil minerals with natural organics and microbes. Soil Sci. Soc. Am. Spec. Publ. 17. SSSA, Madison, WI.

Mohr, E.C.J. 1922. Die Grund van Java en Sumatra. J.H. de Bussy, Amsterdam.

Moormann, F.R. 1981. Representative toposequence soils in southern Nigeria and their pedology. p. 10–29. *In* D.J. Greenland (ed.) Characterisation of soils. Oxford Univ. Press, Oxford.

Neue, U., and H.W. Scharpenseel. 1984. Gaseous products of the decomposition of organic matter in submerged soils. p. 311–328. *In* Organic matter and rice. Int. Rice Res. Inst., Los Baños, Philippines.

Neue, U., and H.W. Scharpenseel. 1987. Decomposition pattern of ^{14}C labelled rice straw in aerobic and submerged rice soils of the Philippines. Sci. Total Environ. 62:431–434.

Nye, P.H., and D.J. Greenland. 1960. The soil under shifting cultivation. Tech. Commun. 51. Commonwealth Bureau of Soils, Harpenden, UK.

Oades, J.M. 1984. Soil organic matter and structural stability: Mechanisms and implications for management. Plant Soil 76:319-337.

Odell, R.T., J.C. Dijkerman, W. van Vuure, S.W. Melsted, A.H. Beavers, P.M. Sutton, L.T. Kurtz, and R. Miedema. 1974. Characteristics, classification and adaptation of soils in selected areas of Sierra Leone, West Africa. Univ. of Illinois Agric. Exp. Stn. Bull. 748.

Parton, W.J., D.W. Anderson, C.V. Cole, and J.W.B. Stewart. 1983. Nutrient cycling in agricultural ecosystems. Univ. of Georgia Coll. Agric. Exp. Stn. Spec. Publ. 23.

Proctor, J. 1984. Tropical forest litter fall II: The data set. Tropical rain forest. p. 83-113. *In* The Leeds Symp. A.C. Chadwick and S.L. Sutton (ed.) Leeds Phil and Lit Soc., Leeds.

Quantin, P. 1972. Les Andosols. Revue bibliographique des connaissences actuelles. Cah. ORSTOM, Ser. Pedol. 10:273-301.

Rennie, P.J. 1955. Uptake of nutrients by mature forest growth. Plant Soil 7:49-95.

Sanchez, P. 1976. Properties and management of soils in the tropics. John Wiley and Sons, New York.

Sombroek, W.G., and W. Siderius. 1981. Nitosols, a quest for significant diagnostic criteria. p. 11-31. *In* International soil museum, annual report, 1981. Int. Soil Museum, Wageningen, Netherlands.

Tanaka, A.T., Sakuma, N. Okagawa, H. Imai, and S. Ogata. 1984. Agro-ecological condition of the Oxisol-Ultisol area of the Amazon River system. Fac. of Agric., Hakkaido Univ., Sapporo.

Tanaka, A.T., Sakuma, N. Okagawa, H. Imai, K. Ito, S. Ogata, and J. Yamaguchi. 1986. Agro-ecological condition of the Oxisol-Ultisol area of the Amazon River system—Report of a survey of Llanos in Colombia and jungle in Peru. Fac. of Agric., Hokkaido Univ., Sapporo.

Tate, K.R., and B.K.G. Theng. 1980. Organic matter and its interaction with inorganic soil constituents. p. 225-252. *In* B.K.G. Theng (ed.) Soils with variable charge. New Zealand Soc. of Soil Sci. Lower Hutt, NZ.

Tisdall, J.M., and J.M. Oades. 1982. Organic matter and water-stable aggregates in soils. J. Soil Sci. 33:141-164.

UNESCO. 1978. Tropical forest ecosystems. A state-of-knowledge report prepared by UNESCO/UNEP/FAO. Natural Resources Res. XIV. UNESCO, Paris.

Vageler, P. 1930. Verlagsgesellschaft fur Ackerbau, Berlin. (English translation, 1933). An introduction to tropical soils. Macmillan Publ., London.

Van Veen, J.A., and E.A. Paul. 1981. Organic carbon dynamics in grassland soils. I. Background information and computer simulation. Can. J. Soil Sci. 61:185-201.

Virakornphanich, P., S-I. Wada, and K. Wada. 1988. Metal-humus complexes in A horizons of Thai and Korean red and yellow soils. J. Soil Sci. 39:529-538.

Voroney, R.P., J.A. Van Veen, and E.A. Paul. 1981. Organic carbon dynamics in grassland soils. II. Model validation and simulation of long-term effects of cultivation and rainfall erosion. Can. J. Soil Sci. 61:211-224.

Wada, K. 1980. Mineralogical characteristics of Andisols. p. 87-108. *In* B.K.G. Theng (ed.) Soils with variable charge. New Zealand Soc. Soil Sci., Lower Hutt, NZ.

3   Myths and Science about the Chemistry and Fertility of Soils in the Tropics

Pedro A. Sanchez
North Carolina State University
Raleigh, North Carolina

Terry J. Logan
Ohio State University
Columbus, Ohio

In many scientific and popular publications, soils of the tropics are considered to be universally acid, infertile, and often incapable of sustained agricultural production (Gourou, 1966; McNeil, 1964; Goodland & Irwin, 1975; Friedman, 1977; Irion, 1978; Reiss et al., 1980; Jordan, 1985). The soil science literature shows that universal tropical soil infertility is a myth devoid of scientific validity. This myth has generated major misconceptions relevant to current global issues such as rural poverty, land degradation, deforestation, biodiversity, and climate change.

The historical development of this misconception has been recently analyzed by Richter and Babbar (1991) who traced it from the initial explorations in the tropics in the early 19th century (Buchanan, 1807), through the prevalence of broad soil genesis theories during the first half of the 20th century (Sibirtzev, 1914; Jenny, 1941), and finally to the lack of utilization of quantitative data about the diversity and management of soils in the tropics, generated largely during the second half of this century. Richter and Babbar cite telling examples of how major ecological texts still use obsolete concepts about soils, and conclude that the myth is a consequence of a major communications gap between soil scientists and other environmental scientists. Newer books, products of multidisciplinary efforts, put this misconception aside (Leith & Werger, 1989; Coleman et al., 1989).

The myth about universal soil infertility in the tropics is readily counteracted by two kinds of evidence. First, the vast diversity of soils in the tropics (Sanchez & Buol, 1975; Moormann & Van Wambeke, 1978; Drosdoff et al., 1978) which is now systematized according to quantitative soil taxonomy (Soil Survey Staff, 1975), a world soil map (FAO, 1971–1979), and numerous and

---

Copyright © 1992 Soil Science Society of America and American Society of Agronomy, 677 S. Segoe Rd., Madison, WI 53711, USA. *Myths and Science of Soils of the Tropics.* SSSA Special Publication no. 29.

increasingly accurate databases and geographic information systems (Sombroek, 1986). Second, the existence of successful, sustained soil management systems in many ecosystems of the tropics (Coulter, 1972; Pushparajah & Amin, 1977; DeDatta, 1981; Sanchez et al., 1987) plus the overwhelming evidence of sustained increases in per-capita food production in tropical Asia and Latin America (Swaminathan, 1982; FAO, 1986). Such successes, however, are concentrated on those soils of the tropics with superior chemistry and fertility and are certainly not sufficient to overcome world food needs or land resource deterioration in marginal areas. Nevertheless, they are successes of such magnitude that food production has outpaced population growth in two of the three tropical regions of the world.

We discuss the overall misconception about universal tropical soil infertility into three specific areas: soil fertility levels, clay mineralogy, and soil organic matter contents.

SOIL FERTILITY PARAMETERS

Soils are as diverse in tropical regions as they are in temperate regions. All 11 soil orders are found in both regions but their distribution varies (Table 3-1). Acid, low fertility soils meeting the stereotypic concept of "tropical soils" are mainly classified as Oxisols and Ultisols. Due to recent global glaciation, these soils cover only 7% of the temperate region but 43% of the tropics. Consequently, the majority of soils in the tropics (57%) does not fit the stereotype. Generally fertile soils, classified as Alfisols, Mollisols, Vertisols, and Andisols, cover a similar proportion of the temperate region (27%) and of the tropics (24%). Further generalizations at the order level are difficult; it is more relevant to examine specific soil-fertility parameters. It is also more relevant to subdivide the tropics into major agroecological zones.

In this chapter five of these zones are discussed: (i) humid tropics, (ii) semiarid tropics, (iii) acid savannas, (iv) tropical steeplands, and (v) tropical

Table 3-1. Approximate distribution of soil orders in the temperate and tropical regions of the world.

	Tropics†		Temperate region‡	
Soil associations dominated by:	Million ha	%	Million ha	%
Oxisols	833	23	7	--
Ultisols	749	20	598	7
Alfisols	559	15	1231	13
Mollisols	74	2	1026	11
Entisols	574	16	2156	24
Inceptisols	532	14	1015	11
Vertisols	163	5	148	2
Aridisols	87	2	2189	24
Andisols	43	1	101	1
Histosols	36	1	204	2
Spodosols	20	1	458	5

† Latitudes below 23½°.
‡ Latitudes above 23½°.

wetlands. The humid tropics have high and constant temperatures and a dry season of <90 d. The humid tropics are, therefore, defined as areas dominated by udic soil moisture regimes or isohyperthermic and isothermic soil temperature regimes. Their native vegetation is tropical rainforest or semideciduous forest. Deforestation is the main land degradation process involved. The semiarid tropics are characterized by a protracted dry season of 6 to 9 mo duration, with a strong ustic soil moisture regime and isothermic or isohyperthermic soil temperature regimes. Their native vegetation are the drier savanna types (Sahelian and Sudanian in West Africa, or thorny vegetation as in the Brazilian Sertão). The acid savannas are seasonal tropics defined by a strong dry season of 3- to 6-mo duration, acid soils, and native savanna vegetation. The tropical steeplands are simply defined as those regions dominated by slopes >30%, mainly in the mountain regions of the tropics and the wetlands as regions with aquic soil moisture regimes. Major limitations in the tropical steeplands are often physical (shallow rooting depth with low available water contents) rather than fertility (Logan & Cooperband, 1987). Erosion of tropical steeplands may increase the base content of those soils formed on more basic parent materials.

Gross estimates of the extensiveness and importance of the main soil fertility constraints are shown in Table 3-2, based on conversion of the FAO soil database into the fertility capability classification system (Buol et al., 1975; Sanchez et al., 1982). Examination of that table destroys the myth of universal tropical soil infertility, and divides the problem into specific soil fertility constraints.

Low Nutrient Reserves

About 36% of the tropics (1.7 billion ha) is dominated by soils with low nutrient reserves, defined as having <10% weatherable minerals in the sand-and-silt fraction. This constraint identifies highly weathered soils with limited capacity to supply P, K, C, Mg, and S. Soils with low nutrient reserves are more extensive in the humid tropics (66%) and in the acid savannas (55%) but are locally important in the Sahel. It is relevant to note that about two-thirds of soils in the tropics (64%) do not suffer from low nutrient reserves.

Aluminum Toxicity

About one-third of the tropics (1.5 billion ha) has sufficiently strong soil acidity for soluble Al to be toxic for most crop species. This constraint is defined as having more than 60% Al saturation in the top 50 cm of soil. Aluminum toxicity is most prevalent in the humid tropics and acid savannas but occurs in large areas of the tropical steeplands. This constraint is found mainly in soils classified as Oxisols, Ultisols, and Dystropepts, and is highly correlated with low nutrient reserves. As in the previous case, it is relevant to note that Al toxicity does not occur in two-thirds of the tropics.

Table 3-2. Main chemical soil constraints in five agroecological regions of the tropics.

Soil constraint	Humid tropics		Acid savannas		Semiarid tropics		Tropical steeplands		Tropical wetlands		Total	
	\multicolumn{12}{c}{million ha and (%)}											
Low nutrient reserves	929	(64)	287	(55)	166	(16)	279	(26)	193	(16)	803	(6)
Aluminum toxicity	808	(56)	261	(50)	132	(13)	269	(29)	23	(4)	1493	(32)
Acidity without Al toxicity	257	(18)	264	(50)	298	(29)	177	(16)	164	(29)	1160	(25)
High P fixation by Fe oxides	537	(37)	166	(32)	94	(9)	221	(20)	0	(0)	1018	(22)
Low CEC	165	(11)	19	(4)	63	(6)	2	(−)	2	(−)	251	(5)
Calcareous reaction	6	(0)	0	(0)	80	(8)	60	(6)	6	(1)	152	(5)
High soil organic matter	29	(2)	0	(0)	0	(0)	--	(0)	40	(7)	69	(1)
Salinity	8	(1)	0	(0)	20	(2)	--	(0)	38	(7)	66	(1)
High P fixation by allophane	13	(1)	2	(0)	5	(0)	26	(2)	0	(0)	50	(1)
Alkalinity	5	(0)	0	(0)	12	(1)	--	(0)	33	(0)	50	(1)
Total area	1444	(100)	525	(100)	1012	(100)	1086	(100)	571	(100)	4637	(100)

Moderate Soil Acidity

Acid soils with surface pH lower than 5.5 but not Al toxicity occupy one-fourth of the tropics (1.1 billion ha) and are important in all agroecological zones. Although correcting soil acidity by liming might be limited to acid-susceptible crops such as cotton (*Gossypium hirsutum* L.) and alfalfa (*Medicago sativa* L.), this constraint is generally associated with somewhat higher fertilizer requirements for these soils than those with higher pH values.

High Phosphorus Fixation

Clayey soils with iron oxide/clay ratios >0.2 fix large quantities of added P (Buol et al., 1975). This constraint, considered very typical of the tropics, is only found in 22% (about 1 billion ha) of the region. It is more extensive in the humid tropics and acid savannas but is also important in the steeplands. Successful management practices to overcome high P fixation in Oxisols have been developed for the acid savannas in Brazil (Goedert, 1985). Since high P fixation is related to high clay content, most sandy and loamy Ultisols and loamy Oxisols do not fix large quantities of P (Lopes & Cox, 1979). Phosphorus fixation is also important in Andisols because of the presence of allophane. This constraint is important in the steeplands and humid tropics where volcanic soils are found. Management practices are different from those designed to overcome P fixation by sesquioxides.

Low Cation Exchange Capacity

Soils with <4 $cmol_c$ kg^{-1} of effective cation exchange capacity (ECEC) occupy only 250 million ha or about 5% of the tropics. Such low ECEC values in the topsoil indicate limited ability to retain nutrient cations against leaching. This is one of the major components of the universal soil infertility myth, alleging that tropical soils are incapable of retaining nutrients. The limit of 4 $cmol_c$ kg^{-1} of ECEC is equivalent to 7 $cmol_c$ kg^{-1} of CEC at pH 7 or 10 $cmol_c$ kg^{-1} of CEC at pH 8.2, the other two commonly used methods (Buol et al., 1975). The main soils exhibiting such low CEC are sandy Entisols, Spodosols, very sandy Alfisols, and Ultisols and acric great groups of Oxisols. The data in Table 3-2 indicate that the vast majority of the tropics (95%) does not suffer from this problem. This is partially due to the higher-than-expected soil organic matter content of soil in the tropics, which provide a source of CEC. Cation-leaching problems do exist in the tropics like they do in the temperate regions, but not to the degree as commonly described.

In addition, many subsoils of Oxisols, Ultisols, and Andisols exhibit significant anion exchange capacity, which decreases leaching losses of nutrient anions such as nitrates and sulfates (Kinjo & Pratt, 1971). This property is also found in Ultisols of southeastern USA, but seldom elsewhere in the temperate region.

Calcareous Reaction

Soils with pH values above 7.3 and with free $CaCO_3$ within the top 50 cm are often deficient in micronutrients, particularly Fe and Zn (Lopes, 1980). Some also show imbalances between Ca, Mg, and K. Although calcareous soils occupy <5% of the tropics, their relative importance is not reflected in that figure, because they are usually intensively utilized, such as in Central Luzon, Philippines (Neue & Mamaril, 1988) and in the Cauca Valley of Colombia (Blasco & Soto, 1978; Ramírez, 1979).

High Soil Organic Matter

Organic matter levels >30% define organic soils (Histosols) and, unlike widespread beliefs, pose major soil fertility constraints. Organic soils are notoriously deficient in Cu, provide poor support for roots, and many exhibit H toxicity. There are approximately 69 million ha of Histosols and other closely related soils in the tropics, of which half are in southeast Asia (Driessen, 1978). Although organic matter is considered a desirable soil property, too much soil organic matter is definitely not (Sanchez & Miller, 1986).

Salinity and Alkalinity

Sixty-six million ha of the tropics have serious salinity problems, with electrical conductivity >4 and 5 dS m^{-1} in the top 1 m. Fifty million ha are alkaline with more than 15% Na saturation within the top 50 cm. Although each of these constraints occupy <1% of the tropics, they are locally important (Aguilera, 1979; Ponnamperuma & Bandyopadhya, 1980). They occur primarily in the humid tropics, semiarid tropics, and wetlands.

Geographical Distribution

The extent of these soil-fertility constraints in the developing world is shown in Table 3-3 for tropical and subtropical areas of Latin America, Africa, and Asia. Their relative importance varies with continental constraints. For example, problems related to soil acidity are more extensive in Latin America than in Africa or Asia. Examination of Tables 3-2 and 3-3 give little support to the myth of universal tropical soil infertility.

CLAY MINERALOGY

A second major myth about the chemistry and fertility of tropical soils is that they are dominated by kaolinite, iron oxides, and other highly weathered clay minerals. Such minerals are now classified as variable charge clays, where CEC increases with soil pH, as opposed to permanent charge minerals such as smectites, illite, vermiculite, and chlorites (Theng, 1980).

Table 3-3. Main chemical soil constraints in the developing world by geographical region (includes tropical and subtropical regions).

Soil constraint	Latin America		Africa		South and southeast Asia		Developing world	
	million ha and (%)							
Low nutrient reserves	941	(43)	615	(20)	261	(16)	1817	(27)
Aluminum toxicity	821	(38)	479	(16)	236	(15)	1527	(23)
High P fixation by Fe oxides	615	(28)	205	(7)	192	(12)	1012	(15)
Acidity without Al toxicity	313	(14)	471	(16)	320	(20)	1104	(16)
Calcareous reaction	96	(4)	332	(11)	360	(23)	788	(12)
Low CEC	118	(5)	397	(13)	67	(4)	582	(9)
Salinity	62	(3)	75	(3)	97	(6)	234	(3)
Alkalinity	35	(2)	18	(--)	7	(--)	60	(1)
High P fixation by allophane	44	(2)	5	(--)	7	(--)	56	(1)
High soil organic matter	9	(--)	12	(--)	23	(1)	44	(--)
Total area	2172	(100)	3011	(100)	1575	(100)	6758	(100)

Uehara and Gillman (1981) calculated the distribution of soils with variable, mixed, or permanent charge. Given the fact that mixtures of clay minerals is the norm rather than the exception in soils anywhere, Gillman and Uehara defined variable-charge soils as those where more than 60% of the ECEC is variable, mixed as those between 40 and 60%, and permanent as those where < 40% of the charge is variable. Also given the fact that the ECEC of soil organic matter is entirely variable, all soils exhibit some degree of variable charge. Their data, shown in Table 3-4, indicate that 60% of soils in the tropics have variable charge, while only 10% have permanent charge. In contrast, 10% of the temperate region soils have variable charge while 45% has permanent charge. There is no question, therefore, that variable charge is the dominant feature of soils of the tropics, but it is certainly not a universal characteristic of the tropics. The 820 million ha of soils with variable charge in the temperate region are found in high latitude Spodosols and Histosols, but also exist in large Ultisol areas such as southeastern USA and southeastern China, and in Andisol regions of Japan, Alaska, the Pacific Northwest, and New Zealand.

Table 3-4. Extent of soils classified as variable, mixed, or permanent charge. Source: Uehara and Gillman (1981).

Soils with:	Tropics		Temperate		World	
	billion ha	%	billion ha	%	billion ha	%
Variable†	3.00	60	0.82	10	3.82	29
Mixed charge‡	1.50	30	3.68	45	5.18	39
Permanent charge§	0.50	10	3.68	45	4.18	32

† 60% of the CEC is variable.
‡ 40-60% of ECEC is variable.
§ <40% of CEC is variable.

SOIL ORGANIC MATTER CONTENTS

It is commonly believed that soils of the tropics have lower organic matter (SOM) contents than soils of the temperate region (McNeil, 1964; Gourou, 1966; Bartholomew, 1972; Jordan, 1985). The red color of many soils in the tropics, high temperatures, and high rainfall are among the reasons cited in support of this generalization. This assumption, however, happens to be wrong. There are no major differences in SOM contents between the two regions (Sanchez & Buol, 1975; Lathwell & Bouldin, 1981; Sanchez & Miller, 1986). Interest in the world C cycle has resulted in many recent estimates of soil C reserves in the humid tropics to answer the question of whether tropical forests are a source or sink of C for the world. Empirical estimates of the soil organic C reserves in the tropics, such as those developed by Schlesinger (1979) and Brown and Lugo (1980), vary widely. Reserves are calculated by multiplying organic C data from a few soil profiles by the land area represented by such soils or by the land area of the ecological region in which they are found. Schlesinger and Brown and Lugo recognized these weaknesses and urged some precise estimates. The soil science literature, however, has many reports indicating that soils of the tropics are not generally low in organic matter. Kellogg (1950) made this point very clear more than 30 yr ago. Studies of several hundred topsoil samples from Hawaii (Dean, 1930), Puerto Rico (Smith et al., 1951), and East Africa (Birch & Friend, 1956) showed average topsoil values on the order of 2% C, a figure that compares favorably with temperate region contents.

Studies involving large numbers of pedons (Post et al., 1982; Zinke et al., 1984; Sanchez et al., 1982) now confirm the prevalent view that soil organic matter contents vary equally in temperate and tropical regions (Duxbury et al., 1989). For example, data from 61 randomly chosen profiles from the tropics and 45 from temperate regions and classified as Oxisols, Mollisols, Alfisols, and Ultisols, showed no significant differences in total C and C/N ratios between soils from tropical or temperate regions at depth intervals up to 100 cm (Table 3-5). Total N contents, however, were significantly higher in the tropical samples while the coefficients of variability were similar. No significant differences in SOM contents were observed between Alfisols from the tropics vs. Alfisols of the temperate region, Ultisols from the tropics vs. Ultisols of the temperate region and Mollisols of the tropics vs. Mollisols of the temperate region (Sanchez et al., 1982). Furthermore, no significant differences in organic matter contents were found between the classic black Mollisols or Chernozems of the temperate region and the red, highly weathered Oxisols of the tropics (Table 3-6).

A recent analysis of well-classified pedons from the National Soil Survey Laboratory database with 282 pedons from the tropics (having isotemperature regimes) and 486 pedons from the temperate regions (non isotemperature regimes) suggest that soils from the tropics may have significantly higher SOM contents to 30-cm depth than soils from temperate regions (Buol et al., 1990). The data shown in Fig. 3-1 is arranged according to the mean annual temperature of the four main soil temperature regimes in soil

Table 3-5. Mean organic matter contents in 61 soils from the tropics vs. 45 soils from the temperate region. Source: Sanchez et al. (1982).

Parameters	Depth, cm	Tropical soils	Temperate soils	Significance	CV% Tropics	CV% Temperate
%C	0-15	1.68	1.64	NS	53	64
	0-50	1.10	1.03	NS	57	69
	0-100	0.69	0.62	NS	59	75
%N	0-15	0.153	0.123	*	62	57
	0-50	0.109	0.090	NS	57	57
	0-100	0.078	0.060	**	54	52
C/N ratio	0-15	13.7	13.6	NS	79	35
	0-50	11.3	11.3	NS	46	32
	0-100	9.6	10.0	NS	46	35

*, ** Significant at the 0.05 and 0.01 probability levels, respectively.

Table 3-6. Mean total C and N reserves of soil orders by geographical regions. Source: Sanchez et al. (1982).

Region	Soil order	No. of profiles	Total C 0-15 cm	Total C 0-100 cm	Total N 0-15 cm	Total N 0-100 cm
Tropics	Oxisols	19	3.8a*	11.3a	0.32a	1.13a
	Alfisols	13	2.9a	6.4b	0.29a	0.85b
	Ultisols	18	2.1b	6.4b	0.18c	0.69b
Temperate	Mollisols	21	3.3a	10.1a	0.27a	0.95a
	Alfisols	16	2.8ab	5.8b	0.20bc	0.56c
	Ultisols	8	2.4b	4.2b	0.15c	0.44c

* Numbers followed by the same lower case letter are not significantly different at the $P < 0.05$ level.

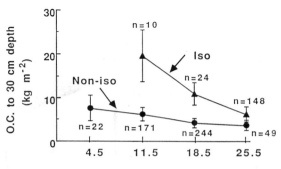

Fig. 3-1. Soil organic C contents as a function of soil temperature regime in iso (tropical) vs. non-iso (temperate) regions. Source: Buol et al. (1990).

taxonomy: frigid (4.5 °C), mesic (11.5 °C), thermic (18.5 °C), and hyperthermic (25.5 °C). Although there is a general decrease in SOM content with increasing temperature, those soils with little seasonal temperature variation indicative of the tropics (isomesic, isothermic, or isohyperthermic regimes) show higher SOM contents than those with strong seasonal temperature variation in mesic, thermic, and hyperthermic regimes.

Soil is a temporary repository for organic C (Buol et al., 1990). Soil organic matter content, therefore, is a function of additions and decomposition rates. Calculations from Greenland and Nye (1959) suggest that generally higher SOM decomposition rates found in the humid tropics caused by high temperatures and ample moisture are balanced by higher litter input, both factors being about five times higher in soils from tropical forests than from temperate forests (Sanchez, 1976). Many processes operating at site-specific rates affect actual input and decomposition rates and provide a wide range of equilibrium SOM contents (Anderson & Swift, 1984). It is safe to assume, therefore, that the range in SOM contents in the tropics is as variable as in the temperate region.

Unfortunately, there is no direct correlation between SOM contents and soil fertility as measured by crop productivity, other factors being constant (Sanchez & Miller, 1986). Mollisols shown in Table 3–6 supported many crops of corn (*Zea mays* L.) without fertilization for years in midwestern USA, Europe, and Argentina (Stevenson, 1984). But many Oxisols with similar SOM contents will definitely not be able to keep one corn plant alive without fertilizer additions (Lopes, 1983). This discrepancy is related to deficiencies of nutrients other than N, and Al toxicity in Oxisols.

CONCLUSIONS

Soils of the tropics are variable in their chemistry and fertility, ranging from the most fertile to the most infertile in the world. Although there is a larger proportion of acid soils in the tropics, they are not even in the majority. The chemical processes involved are the same regardless of latitude. What is different is their management, because of different climate, crop species, and socioeconomic conditions found in the tropics.

Given the expanding databases available from the tropics on inherent soil characteristics, and on their management, it is no longer acceptable to continue the myths of the past that were understandably caused by a lack of systematic study. Site-specific management will be increasingly necessary as demands grow on soils of the tropics for food and fiber production.

REFERENCES

Aguilera, F. 1979. El problema de la salinidad y el sodio en el Valle del Cauca. Suelos Ecuat. 10:98–114.

Anderson, J.M., and M.J. Swift. 1984. Decomposition in tropical rainforests. p. 287–309. *In* S.L. Sutton, et al. (ed.) The tropical rainforest. Blackwell Sci. Publ., Oxford, UK.

Bartholomew, W.V. 1972. Soil nitrogen and organic matter. p. 63-81. Soils of the humid tropics. Natl. Acad. Sci., Washington, DC.

Birch, H.F., and M.T. Friend. 1956. The organic matter and nitrogen status of East African soils. J. Soil Sci. 7:156-167.

Blasco, M., and C. Soto. 1978. Estado actual de las investigaciones sobre micronutrientes en América Latina. Suelos Ecuat. 9:160-164.

Brown, S., and A.E. Lugo. 1980. The role of tropical forests on the world carbon cycle. Centre for Wetlands, Univ. of Florida, Gainesville.

Buchanan, F. 1807. A journey from Madras through the countries of Mysore, Canara, and Malabar. Vol. 2. W. Bulmer and Co., St. James; East India Co., London.

Buol, S.W., P.A. Sanchez, R.B. Cate, and M.A. Granger. 1975. Soil fertility capability classification. p. 126-141. In E. Bornemisza and A. Alvarado (ed.) Soil Management in tropical America. North Carolina State Univ., Raleigh.

Buol, S.W., P.A. Sanchez, J.M. Kimble, and S.B. Weed. 1990. Predicted impact of climate warming on soil properties and use. p. 71-82. In B.A. Kimball et al. (ed.) Impact of carbon dioxide trace gases, and climate change on global agriculture. ASA Spec. Publ. 53. ASA, CSSA, and SSSA, Madison, WI.

Coleman, D.C., J.M. Oades, and G. Uehara (ed.). 1989. Dynamics of soil organic matter in tropical ecosystems. Univ. of Hawaii Press, Honolulu.

Coulter, J.K. 1972. Soil management systems. p. 189-197. In Soils of the humid tropics. Natl. Acad. of Sci., Washington, DC.

Dean, A.L. 1930. Nitrogen and organic matter in Hawaiian pineapple soils. Soil Sci. 30:439-442.

DeDatta, S.K. 1981. Principles and practices of rice production. John Wiley and Sons, New York.

Driessen, P.M. 1978. Peat soils. p. 763-779. In Soils and rice. Int. Rice Res. Inst., Los Baños, Philippines.

Drosdoff, M., R.B. Daniels, and J.J. Nicholaides (ed.) 1978. Diversity of soils in the tropics. ASA Spec. Publ. 34. ASA, Madison, WI.

Duxbury, J.M., M.S. Smith, and J.W. Doran. 1989. Soil organic matter as a source and a sink of plant nutrients. p. 33-67. In D.C. Coleman et al. (ed.) Dynamics of soil organic matter in tropical ecosystems. Univ. of Hawaii Press, Honolulu.

Food and Agriculture Organization of the United Nations. 1971-1979. Soil map of the world. 9 volumes. UNESCO, Paris.

Food and Agriculture Organization of the United Nations. 1986. Yearbook of agriculture for 1985. FAO, Rome.

Friedman, I. 1977. The Amazon basin, another Sahel? Science 197:7.

Goedert, W.J. (ed.) 1985. Solos dos cerrados: Tecnologias e manejo. Nobel, São Paulo.

Goodland, R.J.A., and H.S. Irwin. 1975. Amazon jungle: Green hell to Red Desert? Elsevier, Amsterdam.

Gourou, P. 1966. The tropical world: Its social and economic conditions and its future status. 4th ed. (Trans. by S.H. Beaver and E.D. Laborde.) Longmans, London.

Greenland, D.J., and P.H. Nye. 1959. Increases in carbon and nitrogen contents of tropical soils under natural fallows. J. Soil Sci. 9:284-299.

Irion, G. 1978. Soil infertility in the Amazonian rainforest. Naturwissenschaften 65:515-519.

Jenny, H. 1941. Factors of soil formation. McGraw-Hill, New York.

Jordan, C.F. 1985. Nutrient cycling in tropical forest ecosystems. John Wiley and Sons, New York.

Kellogg, C.E. 1950. Tropical soils. Trans. 4th Int. Congr. Soil Sci. 1:266-276.

Kinjo, T., and P.F. Pratt. 1971. Nitrate adsorption. Soil Sci. Soc. Am. Proc. 35:722-732.

Lathwell, D.J., and D.R. Bouldin. 1981. Soil organic matter and nitrogen behavior in cropped soils. Trop. Agric. (Trinidad) 58:341-348.

Leith, H., and M.H.A. Werger. (ed.). 1989. Tropical rainforest ecosystems: Biogeographical and ecological studies. Ecosystems of the world 14B. Elsevier, Amsterdam.

Logan, T.J., and L.R. Cooperband. 1987. Soil erosion on cultivated steeplands of the humid tropics and subtropics. p. 21-38. In D.D. Southgate and J.F. Disinger (ed.) Sustainable resource development in the Third World. Westview Press, Boulder, CO.

Lopes, A.S. 1980. Micronutrient in soils of the tropics as constraints to food production. p. 277-298. In Priorities for alleviating soil-related constraints to food production in the tropics. Int. Rice. Res. Inst., Los Baños, Philippines.

Lopes, A.S. 1983. Solos sob Cerrado. Inst. da Potassa e Fosfato, Piracicaba, S.P., Brazil.

Lopes, A.S., and F.R. Cox. 1979. Relacão de características físicas, químicas e mineralógicas en solos sob cerrados. Rev. Bras. Cienc. Solo 3:82–88.

McNeil, M. 1964. Laterite soils. Sci. Am. 211:68–73.

Moorman, F.R., and A. Van Wembeke. 1978. The soils of the lowland rainy tropical climates, their inherent limitations for food production and related climatic restraints. Int. Soc. Soil Sci. Trans. 11th Congr. 2:292–312.

Neue, H.E., and C.P. Mamaril. 1985. Zinc, sulfur and other micronutrients in wetland soils, p. 307–320. *In* Wetland soils: Characterization, classification and utilization. Int. Rice. Res. Inst., Los Baños, Philippines.

Ponnamperuma, F.N., and A.K. Bandyopadhya. 1980. Soil salinity as a constraint on food production in the humid tropics. p. 203–216. *In* Priorities for alleviating soil-related constraints of food production in the tropics. Int. Rice Res. Inst., Los Baños, Philippines.

Post, W.M., W.R. Emanuel, P.J. Zinke, and A.G. Stangeberger. 1982. Soil carbon pools and world life zones. Nature (London) 298:156–159.

Pushparajah, E., and L.L. Amin. 1977. Soils under *Hevea* in peninsular Malaysia and their management. Rubber Res. Inst. of Malaysia, Kuala Lumpur.

Ramírez, A. 1979. Evaluacíon de la relacíon calcio/magnesio en suelos del Valle del Cauca. Suelos Ecuat. 10:147–155.

Reiss, S., L. Rother, H. Jensen, B. Came, J. Taylor, and M. Lord. 1980. Vanishing forest. Newsweek 24:117–122.

Richter, D.D., and L.I. Babbar. 1991. Soil diversity in the tropics. Adv. Ecol. Res. 21:315–389.

Sanchez, P.A. 1976. Properties and management of soils in the tropics. John Wiley and Sons, New York.

Sanchez, P.A., and S.W. Buol. 1975. Soils of the tropics and the world food crisis. Science 188:598–603.

Sanchez, P.A., W. Couto, and S.W. Buol. 1982. The fertility capability soil classification system: Interpretation, applicability and modification. Geoderma 27:283–309.

Sanchez, P.A., M.P. Gichuru, and L.B. Katz. 1982. Organic matter in major soils of the tropical and temperate regions. Int. Congr. Soil Sci., 12th (New Delhi) 1:99–114.

Sanchez, P.A., and R.H. Miller. 1986. Organic matter and soil fertility management in acid soils of the tropics. Int. Soc. Soil. Sci. Trans. 13th Congr. 6:609–625.

Sanchez, P.A., E. Pushparajah, and E.R. Stoner (ed.). 1987. Management of acid tropical soils for sustainable agriculture. IBSRAM Proc. No. 2, Bangkok, Thailand.

Schlesinger, W., II. 1979. The role of terrestrial vegetation in the global carbon cycle: Methods I. Appraising changes. G.M. Woodwell (ed.) John Wiley and Sons, New York.

Sibirtzev, N.M. 1949. Soil science. (Transl. N. Kaner). Israel Prog. for Sci. Transl., and USDA, Jerusalem and Washington, DC.

Soil Survey Staff. 1975. Soil taxonomy: A basic system of soil classification for making and interpreting soil surveys. USDA Handb. 436. U.S. Gov. Print. Office, Washington, DC.

Smith, R.M., G. Samuels, and C.F. Cernuda. 1951. Organic matter and nitrogen build-ups in some Puerto Rican soils profiles. Soil Sci. 72:409–427.

Sombroek, W.G. 1986. Establishment of an international soils and land resources information base. SOTER Rep. 1:118–124. Int. Soils Ref. and Info. Centre, Wageningen, Netherlands.

Stevenson, F.J. 1984. Cycles of soil. John Wiley and Sons, New York.

Swaminathan, M.S. 1982. Our greatest challenge: Feeding a hungry world. p. 25–46. *In* G. Bixter and L.W. Shemilt (ed.). Chemistry and world food supplies: The new frontiers. Int. Rice Res. Inst., Los Bannños, Philippines.

Theng, B.K.G. (ed.). 1980. Soils with variable charge. New Zealand Soils Bureau, Lower Hutt, NZ.

Uehara, G., and G. Gillman. 1981. The mineralogy, chemistry and physics of tropical soils with variable charge clays. Westview Press, Boulder, CO.

Zinke, P.J., A.G. Stanganberger, W.M. Post, W.R. Emanuel, and J.S. Olson. 1984. Worldwide organic soil carbon and nitrogen data. ORNL/TM-8875. Oak Ridge Natl. lab., Oak Ridge, TN.

4 Some Aspects of Fertility Associated with the Mineralogy of Highly Weathered Tropical Soils

U. Schwertmann

Institut fur Bodenkunde
Weihenstephan, Germany

A. J. Herbillon

L'Universite de Nancy
Vandoeuvre-Nancy, France

Highly weathered soils of the tropics, such as Oxisols and Ultisols, often have a monotonous mineralogy. Due to their advanced maturity, only the most stable primary and secondary minerals remained. Among these minerals are 1:1 layer silicates of the kaolin group, oxides of Fe, Al, and Ti and some highly resistant minerals either inherited or transformed from the parent rock such as muscovite and Al hydroxy-interlayered vermiculite. The common properties of such soils associated with their mineralogy are low cation exchange capacity (CEC), high zero point of charge (ZPC), high P adsorption and low nutrient reserves combined with high aggregate stability, at least in soils with a high enough clay content (35–40%).

There are indications, however, that these mineral-based properties may, in fact, not be as uniform as is often indicated. It is well known, for example, that Oxisols, although highly matured, still reflect differences in parent rock type (e.g., acid vs. basic igneous rocks) and that yellow low activity clay soils may behave differently from red ones; a difference that is now recognized in the revised key to Oxisols of *Soil Taxonomy* (Soil Survey Staff, 1987).

In this chapter, a few examples are presented to demonstrate that variations *within* the frame of the common Oxisol and Ultisol mineralogy may lead to substantial differences in properties affecting fertility. These differences may be due either to a variation in the proportions of the major minerals or variable characteristics within the same mineral type.

Copyright © 1992 Soil Science Society of America and American Society of Agronomy, 677 S. Segoe Rd., Madison, WI 53711, USA. *Myths and Science of Soils of the Tropics.* SSSA Special Publication no. 29.

1:1 CLAY MINERALS

Properties of Kaolinites

Kaolin minerals from tropical soils show several important differences to kaolinites from kaolin deposits. As shown in Fig. 4-6, kaolinite crystals from such soils may be very small, usually in the range of 0.1 μm, compared to the micrometer-range of "classical" kaolinites. This small particle size gives rise to a much higher surface area. Instead of a few square meters per gram, values of 100 to 250 m^2/g have been reported for Oxisols in South Brazil (Palmieri, 1986). In this case, the crystal plates were only 3 to 8 nm thick and 0.1 to 0.2 nm in diameter. The surface area (S) was related to crystal thickness (CT) by S (m^2/g) = 457 − 33·CT (nm) (Palmieri, 1986) (Fig. 4-1a). The low crystallinity was also related to the x-ray diffraction line intensity ratio $I_{02.11}/I_{001}$, which ranged between 1.5 and 3.0 instead of 1.0 for

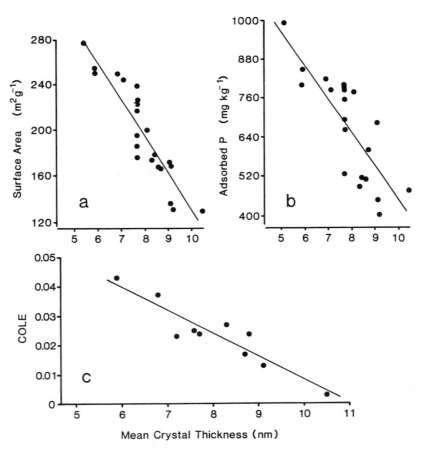

Fig. 4-1. Relation between properties of kaolinites from tropical soils and their mean crystal thickness (Palmieri, 1986).

well-crystallized kaolinite, and in a rather poorly differentiated infrared (IR) spectrum in the OH-stretching range.

The effect of crystal size on soil behavior is substantial. The coefficient of linear expansion (COLE) was found to increase from <0.01 to 0.04 as the crystal thickness decreased from 10 to 6 nm (Fig. 4–1c). In the same range, the phosphate adsorption increased from 400 to 1000 mg kg^{-1} (Fig. 4–1b) (Palmieri, 1986).

Another property characteristic for these small kaolinites is a (although still low) consistently higher structural Fe content than in classical kaolinites. Palmieri (1986) found up to 2% Fe_2O_3, a difference that is in accordance with findings of Mestdagh et al. (1980) for 17 kaolinites from geological deposits vs. five from tropical soils. The latter kaolinites were higher in Fe, less well crystallized, and had a higher surface area (Fig. 4–2). A low crystallinity was also associated with high Fe content in kaolinites from various sources investigated using electron paramagnetic resonance (EPR) by Brindley et al. (1986).

These results suggest that kaolinites formed in an Fe-rich environment will incorporate some Fe into their structure. In turn, it appears to inhibit crystal development and thereby keeps the kaolinite more "active."

The close association between kaolinite crystallinity and Fe content on the one hand and soil environment conditions on the other also evolves from

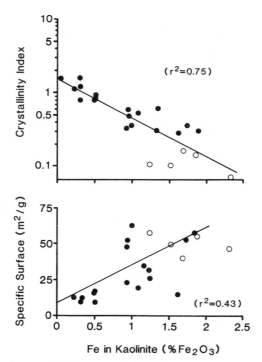

Fig. 4–2. Relation between the Fe content in kaolinites and its crystallinity and surface area (Mestdagh et al., 1980).

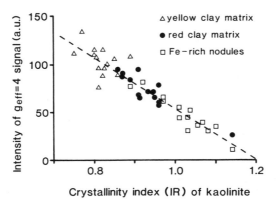

Fig. 4-3. Systematic variation of kaolinite crystallinity in various materials within a deep lateritic profile in Cameroon. Crystallinity was measured by IR absorption and electron spin resonance (Muller & Boquier, 1987).

a study by Muller and Bocquier (1987) of a deep lateritic profile on basement rocks in central Cameroon. In this profile, kaolinites of different compartments (red and yellow clay matrix and red mottles) vary in their Fe content (expressed by the EPR signal at $g_{eff} = 4$) and their crystallinity (as measured from IR absorption), which were again related (Fig. 4-3).

1:1 to 2:1 Mixed Layer Minerals

Delvaux et al. (1990b) conducted a detailed study of the 1:1 clay minerals from an Andept → Tropept → Udalf → Udult pedosequence on basaltic pyroclasts in Cameroon and revealed that (i) the ratio halloysite/halloysite + kaolinite (Ht/Ht + Kt) decreased from 0.98 in the Andept to 0.38 in the more weathered Udult and (ii) that these 1:1 minerals appear to have a substantial proportion of smectite layers intermixed with the dominating halloysite layers, as depicted in Fig. 4-4. This "impure" 1:1 mineral differs

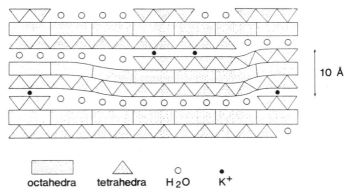

Fig. 4-4. Sketch of the structure of a 2:1 smectite-halloysite mixed layer mineral found in volcanic ash soils in Cameroon (Delvaux et al., 1990b).

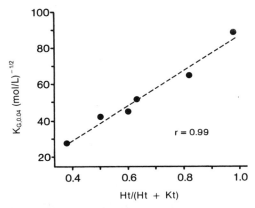

Fig. 4-5. Relation between the Gapon coefficient for K-Ca and the proportion of halloysite in volcanic ash soils of Cameroon (Delvaux et al., 1990a).

markedly in several properties from a pure 1:1 mineral. It has a much higher CEC (at pH 6) of 150 to 370 $mmol_c\ kg^{-1}$, higher Fe and Mg contents of 4 to 7% Fe_2O_3 and 0.3 to 1.0% MgO, respectively, a higher specific surface area of 200 to 240 $mg^2\ g^{-1}$, and a stronger preference for K adsorption as indicated by its k_{K-Ca} Gapon coefficient. Figure 4-5 shows that k_{K-Ca} was linearly related with Ht/(Ht + Kt) (Delvaux et al., 1990a).

These findings have a direct relevance to the management of fertilizer applications in this area. For example, in halloysite-rich soils (Tropept and Udalf) where a high K^+ affinity was observed, K^+ fertilizers are not leached. There it is not necessary to split their application. This is in contrast to what is to be recommended in the allophane-rich and halloysite-poor Andepts areas, where a similar K^+ affinity was not observed.

IRON OXIDES

Adsorption of Phosphate and Trace Metals

The second group of minerals dominating in highly weathered, tropical soils are Fe oxides. Among these oxides, goethite (Gt) and hematite (Hm) are the most important. Their total amounts vary greatly (a few to several 10%). The ratio between the two minerals (Gt/[Gt + Hm] ~ 0.01 − 1.0) also varies. Both oxides are small, with a crystal size of 10 to 50 nm. Therefore, they contribute appreciably to the surface area of tropical soils by 50 to 200 m^2/g of oxide. The small crystals are often poorly developed and heavily aggregated. The two minerals often cannot be distinguished morphologically (Fig. 4-6).

The best-known effect of the Fe-oxide surface is its high affinity towards phosphate retention, as demonstrated by a drastic decrease in P adsorption after differential removal of the Fe oxides by citrate-bicarbonate-dithionite

Fig. 4-6. Electron micrographs of a kaolinite (*upper*), a goethite (*middle*), and a hematite (*lower*) from Oxisols in South Brazil. The Fe oxides were concentrated by a NaOH treatment. Bar represents 100 nm (N Kämpf, Univ. of South Brazil, 1982, unpublished data).

(CBD). A similarly high affinity also exists towards trace metals. Anand and Gilkes (1987) have demonstrated that high percentages of Co and Ni are present in Fe oxides in Australian lateritic soils (Fig. 4-7). Appreciable

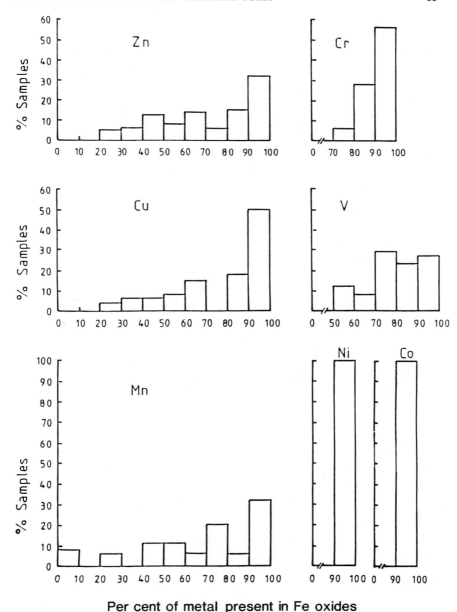

Fig. 4-7. Association of various trace elements with Fe oxides in lateritic soils of western Australia (Anand & Gilkes, 1987).

amounts of Cr, V, Cu, Zn, and Mn are present. Geochemical studies have often revealed significant relations between the content of Fe and these trace elements.

Goethite and hematite have different properties but the properties may also vary for each. With regard to anion and cation adsorption, it is usually assumed that in an aqueous system the surface of both minerals is uniformly hydroxylated (Fe-OH) or hydrated (Fe-OH$_2$) and, therefore, the same for goethite and hematite. On the other hand, IR studies have shown that singly (rather than doubly, Fe$_2$-OH, or triply, Fe$_3$-OH) coordinated Fe-OH groups are by far the most reactive functional groups to form surface complexes with various anions and cations.

To date, studies of this kind have only been made with synthetic hematites and goethites but not with soil Fe oxides. They indicate that the different crystal faces of goethite and hematites adsorb different amounts of phosphate because the densities of Fe-OH groups at different crystal faces vary markedly. Correspondingly, Barrón et al. (1988), who used 43 synthetic hematites with specific surface areas of 6 to 115 m^2 g^{-1}, only found a very weak correlation (R^2 = 23%) between the P adsorption maximum (P$_{max}$) (4.0-146.5 mmol P kg^{-1}) and the surface area. Instead, P$_{max}$ varied strongly between 0.19 and 3.33 μmol P m^{-2}. This variation can be ascribed to the above-mentioned variation in the density of Fe-OH functional groups in conjunction with a varying relative contribution of different crystal faces to the total surface. In brief, only the nonbasal surfaces of hematite (110, 100, and 223) carry Fe-OH groups. The basal surfaces (001, 108, and 104) do not and can, therefore, be expected to be essentially inert, at least at low P concentration in solution.

With goethites, the situation is quite different (Torrent et al., 1991). Here a highly significant relationship was found between P$_{max}$ and surface area for 31 synthetic goethites (surface area 21-115 m^2/g). The slope of this relation gave a value of 2.51 μmol m^{-2} with little variation ($\pm$ 0.17 μmol m^{-2}). This low variation was surprising in view of the fact that the crystals varied greatly in shape. It is, therefore, concluded that the crystals are essentially bound by identical faces, and a value of 2.50 μmol P^{-2} can be computed for the 110 face only, assuming that each phosphate molecule forms a bidentate complex with two neighboring Fe-OH groups. High resolution electron microscopy has indeed shown for synthetic (Schwertmann, 1984) and natural goethites (mineral specimen: Smith & Eggleton, 1983; lateritic soil: Amouric et al., 1986) that 110 faces are essentially the only faces occurring.

It remains to be seen, however, if the much higher P adsorption of 2.51 μmol m^{-2} by synthetic goethites compared to 0.97 μmol m^{-2} by synthetic hematites is of relevance to soil goethites and hematites. Phosphorus adsorption studies with some Ultisols from the southeastern USA and Oxisols from Brazil (Bigham et al., 1978) demonstrated an increase in adsorbed P (mg/g) as the goethite/hematite ratio increased. This may, however, also be due to a higher specific surface area of goethite compared to that of hematite.

Isomorphous Substitution of Iron by Other Trivalent Cations

Iron oxides in soils are hardly ever chemically pure. Because of similar ionic size, the central cation Fe^{3+} can be easily replaced by Al^{3+}, a cation usually present in the pedoenvironment where the Fe oxides form. Although known for long, it has only been recently realized that Al-for-Fe substitution seems to be the rule rather than the exception, and that the degree of substitution in highly weathered soils may reach the maximum possible extent. Its variation has been used to characterize the present and former Al availability in the pedoenvironment, which is determined mainly by pH and degree of desilification. Soil maturity, position in the soil landscape (toposequences), and position in the profile are, therefore, parameters causing Al substitution to vary systematically (Fitzpatrick & Schwertmann, 1982). For example, in a deep laterite profile Muller and Bocquier (1987) found the Al-for-Fe substitution of goethite to increase in the order Fe-rich nodules < red clay matrix < yellow clay matrix (Fig. 4-8). Low substitutions in goethites (0-10 mol%) were found in hydromorphic environments, whereas high values (~33 mol%) occurred in association with gibbsite (Schwertmann & Kämpf, 1985).

To date, it can only be speculated that soil fertility properties may be influenced by Al-for-Fe substitution in goethite and hematite. There may be an indirect effect through crystal size, which has been found to decrease with increasing Al in lateritic soils (Anand & Gilkes, 1987). Again, the properties of synthetic goethites and hematites can be modified substantially by Al substitution. Whereas, Ainsworth and Sumner (1985) found a higher P adsorption per unit of surface area for Al-substituted goethites than for their pure counterparts, Torrent et al. (1990) did not notice any effect of Al. For synthetic hematites, Barrón et al. (1988) noticed a decrease in P adsorption as Al substitution increased.

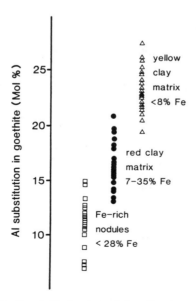

Fig. 4-8. Aluminum-for-Fe substitution in goethites of various materials within a deep lateritic profile of Cameroon (Muller & Boquier, 1987).

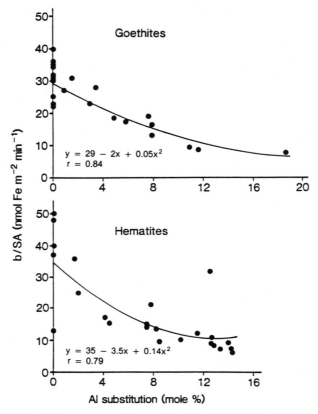

Fig. 4-9. Initial dissolution rate per unit area of surface (b/SA) of synthetic Al-substituted goethites and hematites as a function of the degree of substitution (Torrent et al., 1987).

Thermodynamically, Al-substituted forms should be more stable, as was theoretically derived by Trolard and Tardy (1987). The rate of reductive dissolution of the two minerals in CBD decrease with increasing Al substitution (Fig. 4-9; Torrent et al., 1987), and so did the dissolution rate in a strong acid (HCl) (Schwertmann, 1984). Figure 4-10 compares the dissolution rate of four synthetic goethites with similar substitutions of Fe by Al, Mn, and Cr to that of pure goethite of similar surface area (~ 40 m^2/g). Whereas Al and particularly Cr stabilized the goethite, Mn^{3+} made it more vulnerable to proton attack. Aluminum has been shown to stabilize goethite against transformation to maghemite or hematite by fires, which often occur in tropical areas (Schwertmann & Fechter, 1984).

Soil Mineralogy and Soil Structure in Red-Yellow Topo-sequences

We now have many reports of the occurrence, in the intertropical areas, of chromotoposequences that show a progressive transition from red Ox-

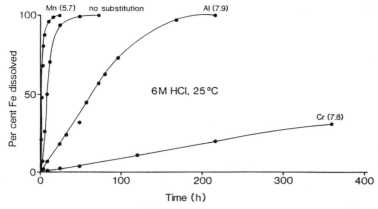

Fig. 4-10. Dissolution-time curves in 6 M HCl of pure, Mn-, Al-, and Cr-substituted synthetic goethites. The figures in parentheses indicate degree of substitution in mol % (Schwertmann, 1991).

isols upslope to yellow Ultisols at midslope (Curi & Franzmeier, 1984; Santana, 1984; Karim & Adams, 1984). Recently, Fritsch et al. (1989) showed that the structure of the top horizons of such a sequence in Ivory Coast drastically varied with the mineralogy. The 2.5 YR Oxisols have a stable, fine granular structure, the 7.5 YR Ultisols show a massive structure, whereas single grain structure predominates in the hydromorphic 10 YR Entisols in the wet lower slope. In the same direction, the clay content (mainly kaolinite) drops from >35% to <5%. Similarly, the Ultisols showed a stronger tendency to surface crusting than the Oxisols when exposed to rainfall.

Although not yet fully understood, these results point to a specific aggregate effect of hematite, the concentration of which drops from 5.6 to 0% along the slope, but may also be associated with the decrease in clay content. Such a situation deserves attention in view of its relevance to soil conservation in regions where structural deteriorations and erosion may lead to harmful degradation of the soil resources.

CONCLUSIONS

Although highly weathered tropical soils generally have a relatively monotonous mineralogy (1:1 layer silicates, Fe oxides, resistant minerals), a closer examination of the 1:1 and Fe-oxide minerals reveal significant differences within each group. Kaolin minerals may differ from classical kaolinites by a smaller particle size, higher surface area, higher CEC and K selectivity, P adsorption, and higher Fe content. Goethite and hematite, as the most widespread Fe oxides in such soils, vary in relative proportion, in crystal size, and in Al-for-Fe substitution. As seen mainly from synthetic samples, these variations can result in differences in surface area, P adsorption, reactivity against protolytic and reductive dissolution, and effects on soil aggregation. Not enough is known about these relations in real soils and research to this end is strongly encouraged.

Dr. E. Murad kindly revised the text.

REFERENCES

Ainsworth, C.C., and M.E. Sumner. 1985. Effect of aluminum substitution in goethite on phophorus adsorption: II. Rate of adsorption. Soil Sci. Soc. Am. J. 49:1149-1153.

Amouric, M., A. Baronnet, D. Nahon, and P. Didier. 1986. Electron microscopic investigations of iron oxyhydroxides and accompanying phases in lateritic iron-crust pisolites. Clays Clay Miner. 34:45-52.

Anand, R.R., and R.J. Gilkes. 1987. Iron oxides in lateritic soils from Western Australia. J. Soil Sci. 38:607-622.

Barrón, V., M. Herruzo, and J. Torrent. 1988. Phosphate adsorption by aluminous hematites of different shapes. Soil Sci. Soc. Am. J. 52:647-651.

Bigham, J.M., D.C. Golden, S.W. Buol, S.B. Weed, and L.H. Bowen. 1978. Iron oxide mineralogy of well-drained Ultisols and Oxisols: II. Influence on color, surface area, and phosphate retention. Soil Sci. Soc. Am. J. 42:825-830.

Brindley, G.W., Chih-Chun Kao, J.L. Harrison, M. Lipsicas, and R. Raythatha. 1986. Relation between structural disorder and other characteristics of kaolinites and dickites. Clays Clay Miner. 34:239-249.

Curi, N., and D.P. Franzmeier. 1984. Toposequence of Oxisols from the central plateau of Brazil. Soil Sci. Soc. Am. J. 48:341-346.

Delvaux, B., A.J. Herbillon, J.E. Dufey, and L. Vielvoye. 1990a. Surface properties and clay mineralogy of hydrated halloysitic clays. I. Existence of interlayer K^+ specific sites. Clay Min. 25:129-139.

Delvaux, B., A.J. Herbillon, L. Vielvoye, and M.M. Mestdagh. 1990b. Surface properties and clay mineralogy of hydrated halloysitic clays. II. Evidence for the presence of halloysite-smectite mixed-layer clays. Clay Min. 25:141-160.

Fitzpatrick, R.W., and U. Schwertmann. 1982. Al-substituted goethite—An indicator of pedogenic and other weathering environments in South Africa. Geoderma 27:335-347.

Fritsch, E., A. Herbillon, E. Jeanroy, P. Pillon, and O. Barres. 1989. Variations mineralogiques et structurales accompagnant le passage "sols rouges-sols jaunes" dans un bassin versant caracteristique du contact foret-savane de l'Afrique Occidentale (Booro-Borotou, Cote d'Ivoire). Sci. Geol. Bull. 42:65-89.

Karim, M.I., and W.A. Adams. 1984. Relationships between sesquioxides, kaolinite, and phosphate sorption in a catena of Oxisols in Malawi. Soil Sci. Soc. Am. J. 48:406-409.

Mestdagh, M.M., Vielvoye, L., and A.J. Herbillon. 1980. Iron in kaolinite: II. The relationship between kaolinite crystallinity and iron content. Clay Min. 15:1-13.

Muller, J.-P., and G. Bocquier. 1987. Textural and mineralogical relationships between ferruginous nodules and surrounding clayey matrices in a laterite from Cameroon. p. 186-194. In Proc. Int. Clay Conf., Denver. 1985. The Clay Mineral Soc., Bloomington, IN.

Palmieri, F. 1986. A study of a climosequence of soils derived from volcanic rock parent material in Santa Catarina and Rio Grande do Sul States, Brazil. Ph.D. thesis. Purdue Univ., Purdue, IN.

Santana, D.P. 1984. Soil formation in a toposequence of Oxisols from Patos de Minas region, Minar Gerais State, Brazil. Ph.D. thesis. Purdue Univ., Purdue, IN.

Schwertmann, U. 1984. The influence of aluminum on iron oxides. IX. Dissolution of Al-goethites in 6 M HCl. Clay Min. 19:9-19.

Schwertmann, U. 1991. Solubility and dissolution of iron oxides. Plant Soil 130:1-25.

Schwertmann, U., and H. Fechter. 1984. The influence of aluminum on iron oxides. XI. Aluminum substituted maghemite in soils and its formation. Soil Sci. Soc. Am. J. 48:1462-1463.

Schwertmann, U., and N. Kämpf. 1985. Properties of goethite and hematite in kaolinitic soils of southern and central Brazil. Soil Sci. 139:344-350.

Smith, K.L., and R.A. Eggleton. 1983. Botryoidal goethite: A transmission electron microscopic study. Clays Clay Miner. 31:392-396.

Soil Survey Staff. 1987. Keys to Soil Taxonomy. SMSS Tech. Monogr. 6. Cornell Univ., New York.

Torrent, J., U. Schwertmann, and V. Barrón. 1987. The reductive dissolution of synthetic goethite and hematite in dithionite. Clay Min. 22:329-337.

Torrent, J., V. Barrón, and U. Schwertmann. 1990. Phosphate adsorption and desorption by goethites differing in crystal morphology. Soil Sci. Soc. Am. J. 54:1007-1012.

Trolard, F., and Y. Tardy. 1987. The stabilities of gibbsite, boehmite, aluminous goethites and aluminous hematites in bauxites, ferricretes and laterites as a function of water activity, temperature and particle size. Geochim. Cosmochim. Acta 51:945-957.

5 Soil Physical Properties of the Tropics: Common Beliefs and Management Restraints

D. K. Cassel

North Carolina State University
Raleigh, North Carolina

R. Lal

Ohio State University
Columbus, Ohio

Soil physical properties have a definite deterministic effect on plant growth and crop production. Because soil physical properties, other than bulk density and soil texture, are difficult, time consuming, and expensive to measure, their importance often receives insufficient attention. Many attempts to establish mechanized management practices on large tracts of land in the tropics have failed. Poor success rate of programs to resettle families in some regions of the tropics have occurred. In part, the failures occurred because sufficient information on soil physical properties was not collected before the projects began. Unfortunately, even when soil physical property information is collected, rarely is it collected in sufficient detail or with appropriate methodology to make optimum land use decisions.

Numerous misconceptions and popular believes have led to inappropriate land use or soil or crop management technologies causing severe soil and water resource degradation. The use of the term *tropical soils* imply uniformity in soil characteristics. The fact is that soils of the tropics are highly variable. Considerable variability in soil physical properties exists in the 1.4 billion ha of land in the tropics (Norse, 1979). In addition, soils of the tropics are believed to be deeply weathered with a well-developed profile amenable to good root system development. Soils in good physical condition are loose, moist, and well aerated with well-connected macropores that allow roots to grow unimpeded. They have a soil temperature warm enough to sustain root growth, but not so warm to slow physiological functioning of the root. Such soils are still found in the near-virgin forests of the tropics, but elsewhere they have been destroyed by the congregation of people, their domestic

---

Copyright © 1992 Soil Science Society of America and American Society of Agronomy, 677 S. Segoe Rd., Madison, WI 53711, USA. *Myths and Science of Soils of the Tropics.* SSSA Special Publication no. 29.

Table 5-1. Extent of selected soil constraints in Tropical America (23 °N–23 °S). (Adapted from Sanchez and Salinas, 1981.)

Soil constraint	Percentage of area†
Low water-holding capacity	42
High erosion hazard	36
Water logging	20
Compaction susceptible	11
Laterite hazard	8

† Land area is 1493×10^6 ha.

animals, and machines as "land is exploited without understanding the relationship between soils, roots, and crop growth" (Trouse, 1979).

Another common belief, based on the high frequency of rains and high relative humidity all year around, is that soil-water reserves are adequate for intensive crop growth. However, shallow-rooted annuals suffer from frequent drought stress because of physical (and chemical) barriers to plant roots, low available water-holding capacities, the presence of plinthite and its formation, rapid decomposition and loss of organic matter, and crusting and soil compaction. Sanchez and Salinas (1981) reported that 42% of the land area in Tropical America has low water-holding capacities (Table 5-1).

Some soils, such as Oxisols and Nitosols, have favorable structure characterized by stable microaggregates. However, a vast majority of soils are characterized by weak structure prone to slaking, crusting, compaction, and a rapid loss of infiltration capacity. Weakly formed structural units slake readily under the impact of high-intensity rains. Consequently, accelerated soil erosion is a severe hazard on most lands with undulating to sloping terrain (Lal, 1984, 1987). The data in Table 5-1 show that at least 36% of the total land area is prone to severe soil erosion. Erosion, both by wind and water, is severe despite the fact that the soil erodibility factor K of most soils is presumably low (Roose, 1977; Lal, 1984). High susceptibility of tropical soils to erosion is related to the predominant effect of raindrop impact leading to high interrill/rill erosion, and mass movement on steep terrains. Above all, the tolerable level of soil loss is drastically low because most plant nutrients are concentrated in the shallow surface horizon and subsoil characteristics are unfavorable for crop growth (Lal, 1987).

Soil temperature regime is another property that is grossly misunderstood. It is commonly believed that intensive cropping, growing two to three crops a year, is possible because of favorable soil temperature regime. In reality, soils in most tropical regions suffer from supra-optimal soil temperatures (Harrison-Murray & Lal, 1979; Maurya & Lal, 1981). Soil temperatures at the 5-cm depth exceeding 45 °C, are common especially in arid and semiarid climates (Lal, 1987). Seedling mortality due to supra-optimal soil temperatures is a principal factor of low crop stand and poor yields.

Soils of the tropics are believed to be deeply weathered with a well-developed profile amenable to good root system development. In reality, however, development of the root system is often severely impaired, and roots of seasonal crops are generally confined to the top layer only. In addition

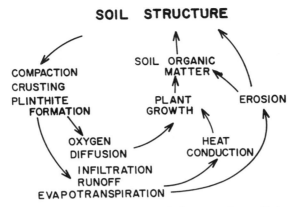

Fig. 5-1. Relationship of soil physical processes and plant growth to soil structure.

to nutrient imbalances and chemical toxicities, root development into the subsoil horizons may also be restricted by physical barriers.

It is because of those and other soil-related constraints that crop yields are often low. The effects of these properties on crop growth is depicted in the simple model in Fig. 5-1. Soil structure of the surface soil and to some degree, the subsurface soil, are affected by soil processes and external forces applied at the soil surface. In turn, soil structure has a direct impact on root growth as well as on the processes of water, solute, air, and heat transport.

The objective of this chapter is to identify major soil physical property constraints to plant growth in the tropics and assess the implication of each constraint on plant growth. Topics include soil variability, soil structural deterioration, low water-holding capacity, soil crusts, soil compaction, plinthite and tillage pan formation, gravel layers, and excessive soil temperature.

SOIL VARIABILITY

Soils located in the large expanse of land in the tropics are highly variable. Soil variability occurs at different levels of observation, ranging from the soil order level to short-range variability occurring within a distance of a few meters. Major soil orders found in the tropics include Alfisols, Aridisols, Entisols, Inceptisols, Oxisols, and Ultisols, but small areas of soils of the remaining four orders are also found (Table 5-2). The variability in soil physical properties arises from differences in vegetation, geology, geomorphology, and rainfall and makes it impossible to generalize about soil physical properties. It is important to realize that processes occurring in soils of the tropics obey the same laws of physics that operate in soils in all regions of the world.

Soils in the tropics have been separated into the following three broad categories: (i) soils containing predominantly low activity clays, (ii) soils with either mixed mineralogy or high activity clays, and (iii) soils of volcanic origin

Table 5-2. Percentage land area of major soil orders in the tropics.†‡

Soil order	Percentage
Alfisol	16
Aridisol	18
Entisol	8
Inceptisol	8
Oxisol	22
Ultisol	11

† Adapted from Buringh (1979).
‡ Total land area in tropics is 4.9 × 10⁹ ha.

(Lal, 1986). In general, the low activity soils have 1:1 clays, mainly kaolinite, and exhibit little shrinking or swelling. Many of the soils in the rainforest and semideciduous forest fall within this category. These soils tend to be deep, highly weathered, have Al and Fe oxides and are classified as Alfisols, Oxisols, and Ultisols (Moorman & Van Wambeke, 1979). Soils with high activities or 2:1 clays occur primarily in arid and semiarid regions. They exhibit a wide range in shrink-swell properties, have stronger soil structure and, therefore, are less easily dispersed. Major soil orders of this group include Aridisols, Inceptisols, and Entisols.

Some appreciation for long-range variability in soil properties can be gained by examining representative soil profiles on a transect across the Amazon. The schematic shows relationhships between geology, landform, and soils (Fig. 5-2). The mineralogy, texture, organic matter content, depth, and degree of development all contribute to differences in soil physical properties. Deep mature soils commonly develop on stable plateaus, table lands, and old sedimentary basins. Younger, less-developed soils occur on recently denuded areas, volcanic areas, and recent sediments (Moorman & Van Wam-

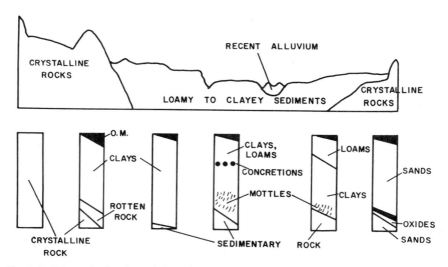

Fig. 5-2. Schematic showing relationship between geology, landform, and soils in a section of Brazil. (Adapted from Sombreck, 1984.)

SOIL PHYSICAL PROPERTIES OF THE TROPICS

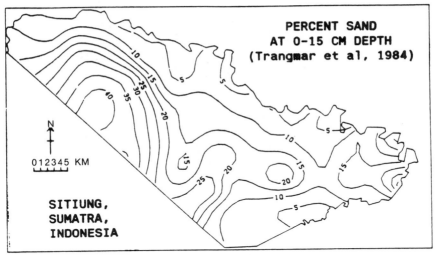

Fig. 5-3. Percent sand at the 0 to 15 cm soil depth at Sitiung, Sumatra, Indonesia. (After Trangmar et al., 1984.)

beke, 1979). The range of properties in soils of the tropics is at least as great as occurs for soils in temperate regions (Moorman & Van Wambeke, 1979).

Variability of soil properties in the 0- to 15-cm depth of soils occurring within an approximately 107 000-ha area in Sitiung, Sumatra, Indonesia was evaluated by Trangmar et al. (1984). The mean percent sand was 13 and ranged from 0 to 59; mean percent silt was 30 and ranged from 7 to 69; and mean percent clay was 57 and ranged from 17 to 82. A map of sand content throughout the region, developed using a kriging interpolation technique, is shown in Fig. 5-3. This observed variability in sand content leads to great differences in soil physical properties and processes within this relatively small geographic region.

Finally, variation in physical properties occurs within distances of a few meters or even less. Short-range variability in soil properties arises due to factors such as the presence of termite mounds (Fig. 5-4 and 5-5), earthworms and their burrowing and casting activity (Fig. 5-6), tree growth, compaction, animal burrows, and tree roots. Some trees have a highly favorable effect on physical properties of soils under its canopy drip. Small lateral variations in texture, organic C, and coarse fragments of gravel or plinthite can have a strong effect on crop performance.

SOIL STRUCTURE

The conventional definition of soil structure as "the arrangement of primary particles into aggregates" is not comprehensive enough to reflect all attributes that affect soil processes and plant growth. An edaphological definition must consider "those properties of soil that regulate and reflect

Fig. 5-4. Maize (*Zea mays* L.) does not grow on the abandoned site of a termite mound.

a continuous array of various sizes of interconnected pores, their stability and durability, capacity to retain and transmit fluids, and ability to supply water and nutrients for supporting active root growth and development."

Considering this as a workable definition, there are many contradictions and conflicting views in relation to structure of soils of the tropics. Most of these contradictions arise due to overgeneralizations of the local specific data pertaining to soil of a limited geographical distribution. A common contradiction lies in the belief that soils of the tropics are extremely well structured, with stable aggregates resistant to compaction and accelerated erosion. The good soil structure is caused by cementation of clay by organo-mineral complexes, especially those of Fe and Al oxides. The stable microaggregates are within the size range of silt to fine sand. Formation of stable microaggregates is related to surface charge properties and amounts of oxides and hydrous oxides of Fe and Al. In these soils, the highest percentage of microaggregates or "pseudo-silt" generally occurs in the argillic horizon. Microaggregation is especially prevalent in Nitosols and Oxisols. That is why their field texture is often loamy rather than clayey as determined by laboratory analy-

Fig. 5-5. Large *Macrotermes* mounds in savanna region, near Mokwa, Nigeria.

sis (Ahn, 1979). Even after treatment with H_2O_2 and prolonged shaking, the aggregates are usually not destroyed. The microaggregates are broken into primary particles only after the addition of an anionic-dispersing agent. While this is true of some soils of volcanic origin (Andosols), and for those that are Oxisols, it is not the case in other soils. There are many examples of the

Fig. 5-6. Freshly formed casts of *Hyperiodrilus africanus* in western Nigeria. The casting rate under ideal conditions can be 100 t ha^{-1} yr^{-1}.

problems caused by structural collapse, especially under intensive land use for food crop production (Lal, 1985). In fact, surface horizons of many soils, with relatively less clay and organic matter contents, are less aggregated than soils of the temperate zone. In these coarse-textured soils, whatever structural aggregates exist are due mainly to the biotic activity of soil fauna (e.g., earthworm). This is particularly true for soils of the humid and subhumid regions of West and Central Africa.

Because of the variations of structural attributes, and the wide range of processes involved, methods of assessment are also numerous and seemingly contradictory. An appropriate technique for determination of soil structure essentially depends on the processes involved in structural development and objectives. Evaluation of stability of structural aggregates to raindrop impact and proportion of macropores and their continuity are important criteria to evaluate structure for soils prone to erosion by water; on the other hand, porosity and pore-size distribution are important for root growth and development. Evaluation of total porosity and pore-size distribution are difficult, however. Porosity is a transient property, and difficult to characterize. Furthermore, critical limits of porosity differ among soils. Some soils are naturally well aggregated whereas others are easily compacted by methods of land development and seedbed preparation. Quite often, the methods used to evaluate pore-size distribution alter the characteristics that are to be determined.

MECHANICAL IMPEDANCE

Mechanical impedance is the mechanical resistance the soil offers to shoot emergence and root growth. Mechanical impedance is intimately related to soil structure. A favorable soil structure for root penetration is one that has a wide range of pore sizes. In addition, the individual structural units are somewhat resistant to external forces that tend to force soil particles closer together.

Compaction, crusting, and pan formation are three densification processes that occur in many soils. These processes create soil conditions with increased mechanical impedance and often restrict rooting activity. *Compaction* is the process in which aggregates or peds are destroyed and individual soil particles are pushed closer together in response to external forces such as rainfall impact, wheeling, and foot traffic. Those soils with greater structural stability are more resistant to compaction. Surface crusting results when soil aggregates at the soil surface are dispersed, often due to high kinetic energy associated with intense rainfall events. Upon drying, the dispersed particles form a thin, dense layer, through which shoots often have a difficult time emerging (Fig. 5-7). Crust formation is more common on soils having weak structure, lower organic matter, and high silt content. Pan formation includes the formation of plinthite or iron pan (Fig. 5-8 and 5-9) (Moorman & Van Wambeke, 1979), but also includes the formation of tillage-induced pans (Alegre et al., 1986b).

SOIL PHYSICAL PROPERTIES OF THE TROPICS

Fig. 5-7. Soils low in organic matter content and containing predominantly low activity clays develop surface seals or crusts that inhibit germination.

Each of these three densification processes reduces total soil porosity at the expense of reducing the volume of macropores, and increases bulk density and soil strength. Root penetration, however, does not depend upon bulk density alone. Resistance to root proliferation is a function of mechanical impedance or shear strength that is affected by both bulk density and soil matric potential (or soil water content). The relationship among soil resistance, bulk density, and matric potential are illustrated for Apomu loamy sand (Paleustalfs) (Maurya & Lal, 1979) (Fig. 5-10). In general, resistance of the soil to a penetrometer increases with an increase in bulk density or a decrease in water content. For Apomu soil, the relative effects of bulk density on penetrometer resistance increased as soil matric potential decreased.

Soil Compaction

For Alfisols and Ultisols of subhumid and semiarid regions, aggregates formed under natural vegetation are not stable and collapse easily under intensive arable land use. Favorable structural properties of these soils in the rainforest are due to relatively high organic matter content and an intense activity of soil fauna (e.g., earthworm). With intensive farming, with or without mechanized operations, structural units disintegrate into individual soil separates. Consequently, these soils become prone to compaction, especially with the use of motorized equipment.

Soil compaction is a more severe problem in the tropics than previously thought (Lal, 1986). Soil compaction cannot occur unless the structural units are broken into smaller units. Once this occurs, usually in response

Fig. 5-8. Some soils in the tropics have shallow effective rooting depth due to occurrence of plinthite near the soil surface.

to externally applied forces, some of these fragments are pushed tightly into existing pore space. Compaction of many soils begins with a tillage operation. Walking on a freshly tilled soil, especially when wet, often compacts the soil to a bulk density greater than it had prior to tillage. Even Oxisols, supposedly of stable structure, succumb to excessive traffic. In Bahia, Brazil, Silva (1981) reported that mechanized cultivation decreased total porosity of the 0- to 30-cm depth from a mean of 55 to 27% (Table 5-3).

Soils with kaolinite or other low activity clays shrink and swell very little and thus do not readily regenerate soil structure once the structural units are crushed. Likewise, the loss of organic matter in newly cleared soils leads to conditions more favorable for compaction. All soil compaction in the tropics is not caused by machinery or animals; much of it is caused by people as they till, harvest, plant, or otherwise manage the field. Unfortunately, most human-induced compaction is not even recognized as compaction,

SOIL PHYSICAL PROPERTIES OF THE TROPICS

Fig. 5-9. Crop stand and yield are very poor when hardened plinthite is exposed to the soil surface.

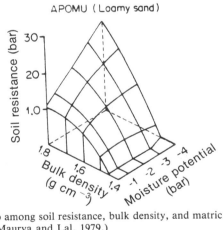

Fig. 5-10. Relationship among soil resistance, bulk density, and matric potential for Apomu loamy sand. (From Maurya and Lal, 1979.)

Table 5-3. Effect of mechanized cultivation on compaction of an Oxisol in Bahia, Brazil (Silva, 1981).

Depth	Without mechanization		With mechanization	
	Bulk density	Porosity	Bulk density	Porosity
cm	g cm^{-3}	%	g cm^{-3}	%
2.5	1.05	58	1.53	14
10.0	1.29	54	1.64	37
20.0	1.30	54	1.81	22
30.0	1.30	55	1.70	36

but it has the same detrimental effects on crop emergence and root development as machinery compaction (Trouse, 1979).

Some tropical soils, especially those containing low soil organic matter and clay contents, set hard due to ultradesiccation. These hard-setting soils (Mullins et al., 1990) are common in subhumid and semiarid tropics. Soils of these regions are easily desiccated and set hard on ultradesiccation when subjected to extreme and prolonged drying. Combined with low organic matter content and predominantly low activity clays, these soils acquire an extremely hard consistency and are often compacted even in their natural state. Hard setting is a severe problem in soils of arid and semiarid regions of Africa, Asia, and Australia.

Root elongation rate is a function of bulk density and soil water content ($g\ g^{-1}$), the same two factors that affect mechanical impedance. The root elongation rate of soybean (*Glycine max* L. Merr.) was less at the same bulk density and water content for Egbeda sandy clay than for Apomu loamy sand (Alfisols) (Fig. 5-11). This result is reasonable for the controlled conditions of the experiment, but in the field situation, a loamy sand soil is typically packed to a higher bulk density than a sandy clay for a given compactive effort. Moreover, loamy sands retain less water than sandy clays. Therefore, under field conditions, the reduction in root elongation for a given compactive effort usually is greater for a loamy sand than for a sandy clay soil.

Much of the data on compaction of soils of the tropics have been associated with mechanical land clearing or reclamation of cleared land (Lal & Cummings, 1979; Seubert et al., 1977; Hulugalle et al., 1984). The data in Table 5-4 show that bulk density increased in the 0- to 15-cm depth of Yurimaguas soil (Ultisol) for all methods used to clear a secondary forest

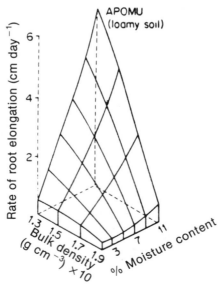

Fig. 5-11. Root elongation rate of soybean as a function of bulk density and soil moisture content of Apomu loamy soil. (From Maurya and Lal, 1979.)

SOIL PHYSICAL PROPERTIES OF THE TROPICS

Table 5-4. Bulk density of Yurimaguas soil before and 14 wk after clearing a secondary forest. (From Alegre et al., 1986a.)

Time	Clearing method	Depth, cm	
		0–15	15–25
		Mg m^{-3}	
Before clearing		1.16b*	1.39b
Fourteen weeks after clearing	Slash	1.27a	1.37b
	Straight blade	1.42a	1.49a
	Shear blade	1.28a	1.50a

* Means within a column followed by the same letter are not significantly different at the 0.05 level.

(Alegre et al., 1986a). The foot traffic and tree felling associated with manual slash clearing increased bulk density from 1.16 to 1.27 Mg m^{-3}. Compaction at the 15- to 25-cm depth did not occur for slash clearing, but did for machine clearing. Mechanical impedance in the 0- to 15-cm depth for the same soil for three management systems was measured with a cone penetrometer 20 wk after clearing (Table 5-5). Foot traffic of farm laborers during hand weeding was a significant factor in compacting the soil surface of the flat planted systems. When the soil was mounded or bedded before planting, soil compaction was less of a problem because workers walked in the furrows rather than on the sloping beds.

Management practices that increase the porosity of compacted sandy and sandy clay soils may improve the rooting patterns of cultivated plants. The increase in porosity not only encourages deeper and denser root growth, but also increases infiltration and water storage, and decreases the distance between roots thus making mineral nutrients more readily available. For sandy soils in general, Nicou and Chopart (1979) found a dramatic increase in root density of rainfed rice (*Oryza sativa* L.) and maize as total pore space increased above 44% by volume (Fig. 5-12). Even though 44% pore space was considered to be high, the pore-size distribution was homogeneous, and individual pores were so small that they created an impedance to root growth. Tillage created larger pores through which the roots penetrated. Large pores to facilitate root penetration can also be created by growing deep-rooted perennials. Hulugalle and Lal (1986) reported that corn following pigeonpea [*Cajanaus cajan* (L.)] had a deeper root system than corn following corn.

Stoner and Freitas (1989) examined the ability of corn, soybean, wheat (*Triticum aestivum* L.) and bean (*Phaseolus vulgaris* L.) to penetrate three soils from the Cerrado of Brazil. Surface material from three soils, identi-

Table 5-5. Log$_{10}$ cone index and water content for the 0- to 15-cm depth of Yurimaguas soil 29 wk after clearing for three soil management systems. (After Alegre et al., 1986a.)

Soil management	Log$_{10}$ CI	Soil water content
	kPa	g g^{-1}
Flat planted	2.681a	0.211
Flat planted, fertilizer, lime	2.635a	0.226
Bedded, fertilizer, lime	2.460b	0.209

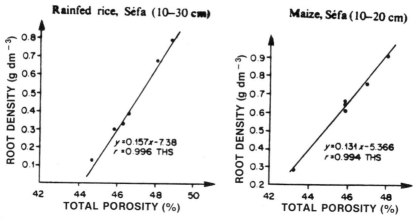

Fig. 5-12. Root density of rainfed rice and maize as a function of total porosity of sandy soil. (From Nicou and Chopart, 1979.)

fied as dark red latosols, were sandy loam (16% clay content), clay (53% clay), and clay (70% clay). These materials were packed to bulk densities of 1.0, 1.1, 1.2, 1.3, and 1.4 g cm^{-3}. Roots of all four crops developed normally at all bulk densities for the sandy loam soil. For the soil with 53% clay, a slight restriction to root penetration occurred for all four crops at the bulk density of 1.4 g cm^{-3}. For the soil with 70% clay, no root development of bean occurred at 1.1 g cm^{-3}; no root development of corn or bean occurred at 1.2 g cm^{-3}; no root development of wheat, corn, or bean occurred at 1.3 g cm^{-3}; and no root development of any crop occurred at 1.4 g cm^{-3}.

Crusting

Many soils upon drying form surface crusts, which is not synonymous to hard setting. Crust strength depends upon physical and chemical factors such as soil texture, soil structure, aggregate stability, organic matter content, and water content (Taylor, 1974). Crusts are more likely to form and impede seedling emergence in soils with weak structure. Crust strength increases as silt content increases and organic matter content decreases. It is not surprising, therefore, that a wide range in behavior with respect to crust formation occurs in tropical soils.

In addition to low organic matter content and the role of clay, soil texture also plays an important role in determining structural characteristics. Silt/clay ratio is an important factor in delineating soil physical properties, and for soil-forming processes such as clay migration. Most soils of the humid and subhumid West African regions, particularly those derived from the basement complex rocks, have low silt content. Consequently, these soils occupy the extreme left-hand corner of the textural diagram (Fig. 5-13). For some soils, with stable microaggregates, the natural texture (feel) can be considerably different from that indicated by their particle-size distribution. The method of mechanical analysis of soil must, therefore, take into considera-

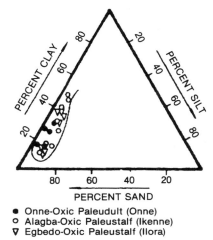

Fig. 5-13. The textural properties of some soils of West Africa with low silt contents.

tion the possible presence of unusually stable silt-sized aggregates. Hossner et al. (1989) identified these stable aggregates in the Sahelian region in West Africa where they identified characteristics of unproductive "ganggani" soils to have a loamy sand texture, a weak structure with a feel much like that of brown sugar, and a surface crust. In general, Oxisols containing kaolinitic clays, hydrous oxides of Fe and Al, and exchangeable Al tend to have stable structure and are less subject to crust formation. Structural stability, however is tempered by organic matter content and texture.

Pans

Several kinds of root-limiting layers occur in tropical soils. In addition to physical constraints, other root-limiting barriers to root growth in soils are high water tables and chemical barriers such as high Al levels (Gonzalez et al., 1979), very low soil P values, or high P-fixing capacities (Sanchez & Salinas, 1981). Some soils, especially those in arid and semiarid regions, are characterized by the presence of a dense pan, even when the clay content exceeds 40%. These soils with naturally occurring pans restrict root development of even deep-rooted perennials. Spodosols have a spodic horizon with high organic matter content. This organic pan occurs below the tillage depth, has high mechanical impedance, and presents a barrier to root growth, whenever roots extend to reach it. These soils usually are not planted to crops. Iron pans or plinthite are natural pans or layers that occur in soil under certain environmental conditions, but their occurrence is less common than previously thought (Sanchez, 1976). Tillage-induced pans are present in some soils in the tropics, but the existence of this pan, although present, often may not be recognized.

Table 5-6. Bulk density of Yurimaguas soil at 15- to 25-cm depth at 8 and 23 mo after clearing and imposing various tillage regimes. (From Alegre et al., 1986a.)

Treatment	Bulk density before clearing	Bulk density after:	
		8 mo	23 mo
		g cm^{-3}	
Slash/burn/no-tillage	1.39	1.37b*	1.38b
Straight blade/rototill	1.39	1.44ab	1.56a
Shear blade/rototill	1.39	1.52a	1.58a

* Entries in a given column followed by the same letter are not significant at the 0.05 level.

Tillage-Induced Pans

Tillage-induced pans, often called *plow pans*, form in coarse-textured soils having single grained or weak soil structure. According to Trouse (1979), plow pans are formed in response to pressure from a wheel riding on soil in the moldboard plow furrow. The wheel compacts tilled soil that was loosened on the previous tractor pass and sluffed into the furrow. This layer of loose soil is compacted, often to a bulk density that impedes root growth. When this process is repeated at approximately the same soil depth for several years, the pan can become quite dense and impenetrable to roots. Tillage pans are also formed in response to other tillage procedures, such as disking or rototilling.

Tillage-induced pans in Ultisols in the Amazon are similar to those formed in Ultisols in temperate regions. Bulk density in the 15- to 25-cm depth of a newly cleared Yurimaguas soil subject to roto-tillage continued to increase with time while bulk density of the soil managed by manual labor did not increase (Table 5-6). Some densification in this layer occurred during the land-clearing process, but pan formation continued in response to structural degradation caused by the rapidly rotating rototiller tines. Deep tillage such as subsoiling or chisel-plowing is effective in breaking up these pans (Fig. 5-14). Under a mechanized cultivation of an Oxisol in Brazil, the soil was compacted to the 30-cm depth and created a tillage pan at the 20-cm depth (Silva, 1981).

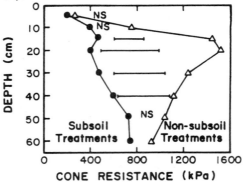

Fig. 5-14. Penetrometer cone resistance vs. soil depth for subsoiled vs. nonsubsoiled tillage treatments on Yurimaguas soil (J.C. Alegre, 1900, unpublished data).

Laterite and Plinthite

For years we have heard that soils of the tropics, if allowed to dry out after clearing, will turn into laterite or ironstone (McNeil, 1964). Laterite is a hard sheet of Fe or Al-rich duricrust. Iron or Al-rich layers result from natural processes that remove silica and accumulate sesquioxides. Within the soil profile, these soil layers are often soft. When exposed and desiccated they harden through Fe-Al cementation into rock-like structure. These hard crusts hinder normal farming operations, restrict root growth, and reduce productivity. About 130 million ha of forested soils and 113 million ha of savanna soils in West Africa have in situ concretionary or Fe pans (Obeng, 1978). Laterized soils are extensive in West Africa and may cover as much as 50% of the land surface in Niger, Burkina Faso, and other regions with a rainfall of 200 to 500 mm yr^{-1}. Many believe the formation of laterites is due to ferruginisation of termite nests and gallery systems. On the contrary, however, the process of laterization is caused by natural physicochemical processes. The development of vesiculae by the removal of soft material is caused by percolating water and not by burrowing termites.

The laterite hazard, although important at specific locations, is a minor problem in Tropical America and most of the plinthite occurs in the subsoil on flat topography with only minor susceptibility to erosion (Buol & Sanchez, 1979; Sanchez & Salinas, 1981). Plinthite is an Fe-rich soil material, having mottles, with a tendency to irreversibly harden into ironstone as a result of alternating soil water saturation and drying conditions. It is estimated that the extent of land areas with plinthite in the tropics is about 7% and is <2% in the Amazon. Plinthite is associated with a fluctuating shallow water table that carries Fe^{2+} to the appropriate depth where it oxidizes, precipitates, and accumulates as Fe^{3+} (Moorman & Van Wambeke, 1979). Where plinthite occurs under humid conditions in the Amazon, it does not occur as massive sheets, but usually as concretions. These concretions have no direct effect on growth other than to reduce the soil volume accessible for root exploration and water and nutrient storage. In some areas, the long-term effect of repeated burning of vegetation has been to change the climax vegetation from forest to savanna. Accompanying changes in water balance and soluble base status are partly responsible for the occurrence of plinthite and hardened ironpans.

Gravel Layers

Some upland soils are characterized by one or more well-developed gravelly horizons at shallow depths (Table 5-7; Fig. 5-15). These gravelly horizons are often compacted and rigid because of the clay matrix, and can cause mechanical impedance to root development of many shallow-rooted crops (Babahola & Lal, 1977; Vine et al., 1975). In addition to the more-resistant weathering material (i.e., quartz) localized concentrations of concretionary materials (Fe and Mn concretions) also occur in soils with a fluctuating water table. Soils with high gravel contents have usually high bulk density ranging

Table 5-7. Gravel content of different horizons for some soils of southwest Nigeria (IITA, 1975).

Series	Location coordinate	Gravels in different horizons, %					
		1	2	3	4	5	6
Egbeda	9°N, 4°E	5	57	56	54	45	37
Sepeteri	9°N, 4°E	11	56	69	74	67	39
Iso	8°N, 3°E	5	34	58	29	18	--
Fashola	8°N, 3°E	5	39	65	69	72	--
Iwaji	8°N, 4°E	29	68	72	62	--	--
Erinoke	8°N, 4°E	0	43	73	63	39	--
Ekiti	7°N, 4°E	3	7	32	85	88	--
Apomu	9°N, 4°E	0	0	0	6	12	23

Fig. 5-15. Quartz gravel, of varying concentration and size, occurs at shallow depths within the soil profile.

between 1.5 and 1.9 Mg m^{-3}. However, the bulk density of the gravel-free material is generally low ranging between 1.1 and 1.4 Mg m^{-3}.

SOIL-WATER RELATIONS

The quantity of water present in the rooting zone of a soil at any given time is a function of the water balance, that is, to the additions and losses of water from the rooting zone of a particular crop and the capacity of soil to store water in the range available for crop growth. Although some land in the tropics is irrigated, most is rainfed. Depending on soil conditions, water arriving at the soil surface may be lost to runoff, or it may infiltrate the soil and be temporarily stored until it is lost to deep or lateral drainage, or to evapotranspiration.

Mean annual rainfall in the tropics ranges from <250 mm to >6000 mm, with annual rainfall in some regions in West Africa and northeast India exceeding 12 000 mm (Lal, 1986). Even though total annual rainfall may be adequate, the variation in its frequency, intenity, and duration can be so great as to lead to periods when plants undergo appreciable moisture stress. Other factors leading to periods of yield-limiting plant water stress are low infiltration rates, excessive drainage, and low water-holding capacities. The rooting depth and water storage capacity in the rooting zone become important during these droughty periods.

The fate of precipitation, once it arrives at the soil surface, is intimately linked to soil structure. If a soil is crusted, compacted, or puddled, its ability to transmit water through the soil surface is low thus promoting surface runoff rather than infiltration. Water lost to runoff from a particular land area is lost to crop use on that area, although the water may be used for crop production on land at a lower position on the landscape. Large amounts of surface runoff often lead to accelerated erosion and subsequent loss of land for crop production (Lal, 1986).

Infiltration

In many cases, the rate at which water infiltrates the soil surface is controlled by the rate at which water (rainfall and irrigation) reaches the soil surface. When water arrives at the soil surface at a rate greater than the maximum rate at which it can infiltrate the soil, then soil properties and the antecedent soil water content control the infiltration process. The structural condition of the soil surface, including pore-size distribution and pore interconnectivity become important. In general, in the rainy tropics, Oxisols of medium to high clay content offer adequate porosity to allow rapid infiltration into the subsoil (Moorman & Van Wambeke, 1979). The infiltration capacity of undisturbed soils of low activity clays under natural vegetation cover in the subhumid and semiarid regions is also high. With cultivation, however, the infiltration rate declines because of surface crusting and reduction in the proportion of macropores. The decline in infiltration occurs regardless of

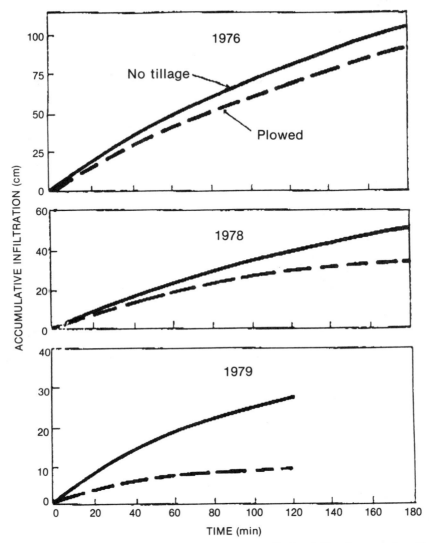

Fig. 5-16. Effects of land use and cropping systems on equilibrium infiltration rate of an Alfisol at IITA, Ibadan, Nigeria.

the farming systems, with motorized tillage operations being a major factor responsible for structural collapse. The rate of decline is, however, less with traditional farming and slash-and-burn methods than with intensive agriculture (Fig. 5-16 and 5-17; Table 5-8). Many of the Ultisols and Alfisols, however, contain large amounts of fine sand and silt, which can lead to crusting or compaction that reduce the infiltration rate.

Infiltration rate measurements are highly variable. It is rare that a sufficient number of measurements are taken to adequately characterize the in-

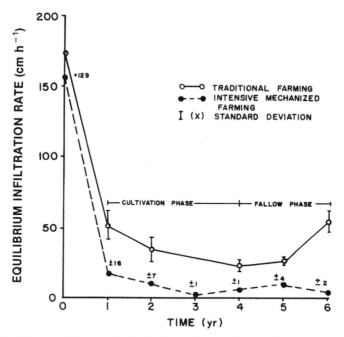

Fig. 5-17. Effects of tillage method and soil application of insecticide on cumulative water infiltration (Lal. 1985).

filtration process. Cumulative infiltration during a 2-h period in a Yurimaguas soil, before clearing a 20-yr-old secondary forest, exceeded 400 mm h^{-1} (Fig. 5-18) (Alegre et al., 1986a). Texture of the A horizon was 63% sand, 10% clay, and 27% silt. Manual clearing did not alter the infiltration rate, but compaction caused by mechanical clearing with both shear and straight blades reduced cumulative infiltration to < 10% of its original value. Chisel plowing and deep disking the mechanically cleared land prior to planting eliminated much of this compaction. Infiltration rate on the manually cleared land 2 yr after clearing was only 25% of its initial value, but was only 3%

Table 5-8. Effects of land-clearing methods on physical properties of an Alfisol in western Nigeria.†

Clearing method	Soil bulk density, 0- to 10-cm depth		Penetrometer resistance		Hydraulic conductivity	
	Initial	Final	Initial	Final	Initial	Final
	— g cm^{-3} —		— kg cm^{-2} —		— cm min^{-1} —	
Mechanical	0.91	1.25	0.50	5.13	16.1	1.29
Slash and burn	0.86	1.12	0.67	2.89	15.2	5.04
Slash	0.89	1.13	0.60	1.86	9.75	4.61
LSD (0.05)	0.29		0.35		9.3	

† Adapted from Lal and Cummings (1979).

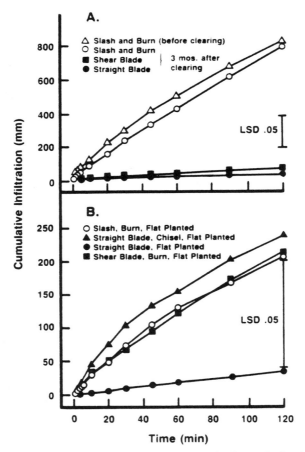

Fig. 5-18. Cumulative infiltration as affected by (A) land-clearing method and (B) soil management system. (After Alegre et al., 1986a.)

of its initial value for land cleared by a straight blade without disking. Chisel plowing and subsoiling can be used to disrupt tillage-induced pans thereby increasing the infiltration rate (Alegre et al., 1986b).

The observations on infiltration for the Yurimaguas soil described above are quite different from those reported for a Typic Haplorthox in Sumatra, Indonesia (Makarim et al., 1988). Even though much of the topsoil had been removed by mechanical land clearing, the infiltration rate was 200 mm h^{-1}. Texture of the soil was 74% clay, 12% silt, and 14% sand. One year after various crop management systems were imposed on this degraded soil, cumulative infiltration actually increased, with higher rates where applications of lime and P were greatest (Table 5-9).

The importance of the infiltration process was recognized by Sanchez et al. (1982) in the development of the Fertility Capability Classification System. The first categorical level in the system is *type* (topsoil texture); the

SOIL PHYSICAL PROPERTIES OF THE TROPICS

Table 5-9. Water infiltrated in 1 h in a Typic Haplorthox in Sumatra as affected by fertility level regime established for a period of 1 yr. (From Makarim et al., 1988.)

Fertility rate	Water infiltrated, mm
Control	256a†
Low	290ab
High	470b

* Means with the same letter are significantly different at the 0.05 level.

Table 5-10. Interpretation of types and substrata types in the Fertility Capability Classification System. (From Sanchez et al., 1982.)

Type	Infiltration rate	Water-holding capacity
S: sandy	High	Low
L: loamy	Medium	Good
C: clayey	Low	Good

second, *substrata type* (subsoil texture); the third, *modifiers*. Interpretation of the types and substrata types are actually based on soil physical properties or processes (Table 5-10).

Water Retention

Soils vary widely in their ability to retain water. The amount of water retained by a soil depends on porosity and pore-size distribution that depend on the degree of soil development, soil structure, soil texture, parent material, organic matter content, biological activity, and soil management. Soil management practices often alter porosity and pore-size distribution on a week-to-week or even a day-to-day basis. Because plants are generally able to extract water from soils at water contents above in situ field capacity to the wilting point, soil water characteristic curves often cover the range from saturation to -1500 kPa (Fig. 5-19). Water retention characteristics, water-

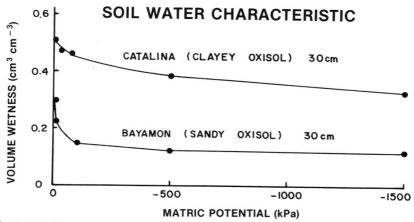

Fig. 5-19. Soil water characteristic for a sandy and a clayey Oxisol in Puerto Rico. (From Wolf and Drosdoff, 1976.)

holding capacity, and available water were reported for four Puerto Rico soils (Wolf & Drosdoff, 1976a,b). In situ field capacity for each soil was measured in the field several days after drainage of well-wetted soil profiles began. Based on tensiometer measurements, the soil water potential at in situ field capacity was found to occur at -7 kPa for sandy Bayamon soil (Psammentic Haplorthox) and -5 kPa for clayey Humatas (Typic Tropohumults), Catalina (Typic Haplorthox), and Torres (Orthoxic Palehumults) soils.

Puddling

The total area of marshes and swamps in the world equals about 2×10^6 km^2, of which 75% is located in the tropics. In the tropics of South and Southeast Asia, swamps have been used intensively and have sustained high population densities for a long period. The objective of seedbed preparation for upland soils are opposite to those of wetlands. In uplands, soil must be tilled only when it promotes structure and improves total and macroporosity. In wetlands, however, the objective is to deliberately destroy structure, decrease total and macroporosity, reduce percolation and leaching, increase waterlogging, and reduce aeration. The process of deliberate destruction of soil structure by physical manipulation or plowing when the soil is at or near the saturation point is called *puddling*. Puddling is done when the soil strength is the lowest.

This seemingly contradictory approach to seedbed preparation for rice paddies vis-a-vis upland crops is partly due to the fact that rice is a semi-aquatic plant. It has a special root system that enables it to grow under an anaerobic environment. Secondly, rice culture evolved as a labor-intensive system. Because transplanting is done manually, it is important that soil is soft enough to facilitate seedling insertion without much hindrance. Thirdly, trafficability, a serious consideration in mechanized farming operations, was not an important issue for small farms of the South and Southeast Asian tropics.

The concept and practices of puddling, contradictory as it may seem, have important economic and environmental implications. The area sown to rice worldwide is expected to increase from 98 million ha in 1975 to about 126 million ha by the Year 2000.

Environmentally, rice paddies are an important source of methane emission into the atmosphere. An average emission of methane from rice paddies during the rice-growing season is estimated to range from 12 to 54 g m^{-2}. The amount of methane emitted depends on soil texture, soil organic matter content, climate, and the degree of puddling and anaerobiosis. The need for puddling, therefore, needs to be addressed for economic, agronomic, and environmental issues.

SOIL ERODIBILITY

Most soils with weakly developed structural units and aggregates are unstable to raindrop impact. These soils slake readily on quick wetting. Con-

Table 5-11. Erosion rates for steep soils in the tropics.

Country	Criteria	Equivalent field erosion rate, Mg ha^{-1} yr^{-1}	Reference
A. Africa			
Ethiopia	Sediment load	165	Virgo & Munro, 1978
Tanzania	Bare soil	38–93	Ngatunga et al., 1984
Ivory Coast	Bare soil	138	Roose, 1975
Ghana	Bare soil	100–313	Bonsu & Obeng, 1979
Nigeria	Bare soil	230	Lal, 1976
Lesotho	Sediment load	180	Chakela, 1981
B. Asia			
India	Crop land gullies	4–43	Dhruva Narayana & Sastry, 1983
Bangladesh	50% slope	520	Islam, 1983
Java, Indonesia	Imperata	345	Suwardjo et al., 1975
C. Tropical America and the Caribbean			
Trinidad	10 to 20° slope, bare	490	Ahmad & Brechner, 1974
Northeast Brazil	Cropped land	115	Ramos & Marinho, 1980
El Salvado	Steeplands	130–260	Wall, 1981
Guatemala	Corn on steeplands	200–3600	Arledge, 1980
Peru	Bare soil	148	Felipe-Morales et al., 1978
Columbia	Crop land	21.5	Rivas, 1983

sequently, soils with rolling to steep relief are severely prone to erosion (Table 5-11). However, susceptibility of soil to water erosion or the erodibility factor K of the Universal Soil Loss Equation (USLE) is often low (Table 5-12). This

Table 5-12. Erodibility of some low activity clay soils determined in field plots in the tropics.

Country	Region	Erodibility	Reference
A. Alfisols			
Benin	Subhumid	0.10	Roose, 1977
Ivory Coast	Subhumid	0.10	Roose, 1977
Kenya	Subhumid	0.03–0.49	Barber et al., 1979
Nigeria	Subhumid	0.06–0.36	Lal, 1976
Nigeria	Subhumid	0.058	Wilkinson, 1975
Tanzania	Semiarid	0.121–0.160	Ngatunga et al., 1984
B. Ultisols			
Hawaii	Humid	0.09	Dangler & El-Swaify, 1976
Nigeria	Humid	0.04	Vanelslance et al., 1984
Thailand	Subhumid	0.09–0.19	Tangtham, 1983
Trinidad	Humid	0.03–0.06	Lindsay & Gumbs, 1982
C. Oxisols			
Costa Rica	Humid	0.103–0.155	Amezquita & Forsythe, 1975
Hawaii	Humid	0.14–0.22	Dangler & El-Swaify, 1976
Ivory Coast	Humid	0.10	Roose, 1977
Puerto Rico	Humid	0.01	Barnett et al., 1971

low erodibility of soils prone to severe erosion is a contradiction created by the definition of erodibility factor K. The factor K of USLE is inversely proportional to the rainfall erosivity (K = A/R, where A is erosion and R is erosivity). Soils in regions of high rainfall erosivity, therefore, tend to have low erodibility and vice versa. If the USLE definition of K is accepted, it seems that soils with low erodibility may be highly prone to erosion.

SOIL HEAT

Soil temperature is an important factor affecting soil physical processes (hard setting, evapotranspiration, soil organic matter accumulation, and biotic activity) and crop growth. In many situations, soil temperature may have a more drastic effect on crop growth than does air temperature. Soil temperature is affected by the energy balance and thermal properties of the soil. The total amount of radiation received at the ground level is an important factor. The total radiation received in the humid tropics is low, and is a severe constraint to the potential maximum yield obtainable in this region.

In subhumid and semiarid regions, soil temperatures as high as 50 °C have been recorded on plowed fields at the 1-cm depth. Consequently, crop establishment failures are common in the West African Sahel and other regions of arid and semiarid tropics. These failures often result from the combined effects of high soil temperature and low soil moisture in the vicinity of germinating seeds and young seedlings.

The soil temperature regime is influenced by and sometimes can be controlled by surface mulches. However, effects of mulching on temperature are highly variable, contradictory, and often confusing. For example, compared to a bare soil, mulch increases soil temperature during cooler weather and in the mornings, and decreases it during hot spells and in the afternoons. The effects on soil temperature also vary among soils, the kind of mulch and rate applied, the antecedent soil moisture content, and the prevalent level of insolation.

The effects of these factors on the soil temperature regime for a West African Alfisol are shown in Fig. 5-20. There are obvious differences in soil temperature measured at the 5-cm depth with respect to mulching, time of the day, method of seedbed preparation, and nature of the mulch material. For example, compared with the flat ground surface, the straw mulch decreased the maximum soil temperature and the clear plastic mulch increased it. The effect of black polythene mulch was similar to that of the ridge treatment. Even the amplitude varied among treatments being 7, 10, 12, 12, and 15 °C for straw mulch, bare soil surface, ridged soil, black polythene, and white polythene mulch, respectively.

The effect of soil temperature on crop growth is also highly variable. In some seasons, decreasing the maximum soil temperature may increase the yield of some crops but not of others. The range of optimum soil temperature depends on the crop, its stage of growth, soil moisture regime, and other factors that affect crop growth and vigor. In addition to the adverse effect

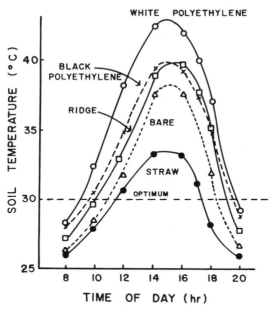

Fig. 5-20. Mean weekly soil temperature at the 5-cm depth on a sunny day in the first growing season of 1976 for different mulch materials (Maurya and Lal, 1979).

of supra-optimal soil temperature on seed germination and seedling establishment, high soil temperatures enhance the rate of organic matter decomposition, facilitate crusting and hard setting, and accelerate laterization. High temperatures also reduce crop growth and yield by curtailing root growth and development, and reducing nutrient and water availability, uptake, and translocation.

CONCLUSION

As the database on physical properties of soils of the tropics is strengthened, some of the popular myths and misconceptions are gradually being replaced by scientific facts. Increased understanding of the processes and properties is a necessary prerequisite to develop sustainable land uses and soil/crop management systems. Because we know that some soils of the tropics have weak structure prone to splash, slaking, crusting, and compaction, we are able to develop management systems to minimize these constraints. Water management systems are developed to increase infiltration capacity or safely dispose excessive runoff during heavy rains, and conserve soil water in the root zone for minimizing water stress during periods of drought. Soil temperature and water regimes can be manipulated through mulch-farming techniques involving conservation tillage, alley cropping, and appropriate crop mixtures. The development of sustainable farming/cropping systems, for high and profitable returns with minimum damage to soils and the environment, depends on the replacement of misconceptions and myths with scientific data.

REFERENCES

Ahn, P.M. 1979. Micro-aggregation in tropical soils: Its measurement and effects on the maintenance of soil productivity. p. 75-86. *In* R. Lal and D.J. Greenland (ed.) Soil physical properties and crop production in the tropics. John Wiley and Sons, New York.

Alegre, J.C., D.K. Cassel, and D.E. Bandy. 1986a. Effects of land clearing and subsequent management on soil physical properties. Soil Sci. Soc. Am. J. 50:1379-1384.

Alegre, J.C., D.K. Cassel, and D.E. Bandy. 1986b. Reclamation of an Ultisol damaged by mechanical land clearing. Soil Sci. Soc. Am. J. 50:1026-1031.

Babalola, O., and R. Lal. 1977. Sub-soil gravel horizon and maize root growth. Plant Soil 46:337-357.

Buol, S.W., and P.A. Sanchez. 1979. Rainy tropical climates: Physical potential, present and improved farming systems. Int. Soc. Soil Sci., 11th Congr. 2:292-312.

Gonzalez-Erico, E., E.J. Kamprath, G.C. Naderman, and W.V. Soares. 1979. Effect of depth of lime incorporation on the growth of corn on an Oxisol of central Brazil. Soil Sci. Soc. Am. J. 43:1155-1158.

Harrison-Murray, R.S., and R. Lal. 1979. High soil temperature and response of maize to mulching in the lowland humid tropics. p. 285-304. *In* R. Lal and D.J. Greenland (ed.) Soil physical properties and crop production in the tropics. John Wiley and Sons, New York.

Hossner, L.R., J.B. Gardiner, M. Doumbia, J. Wendt, and R. Chase. 1989. Causes and control of pronounced plant growth variability over short distances. p. 353-356. *In* Neil Caudel (ed.) Tropsoils Tech. Rep. 1986-1987. North Carolina State Univ., Raleigh.

Hulugalle, N.R., and R. Lal. 1986. Root growth of maize in compacted gravelly tropical Alfisol as affected by rotation with a woody perennial. Field Crops Res. 13:33-44.

Hulugalle, N.R., R. Lal, and C.H.H. ter Kuile. 1984. Soil physical changes and crop root growth following different methods of land clearing in Western Nigeria. Soil Sci. 138:172-179.

Lal, R. 1984. Mechanized tillage systems effects on soil erosion from an Alfisol in watersheds cropped to maize. Soil Tillage Res. 4:349-360.

Lal, R. 1985. Mechanized tillage systems effects on physical properties of an Alfisol in watersheds cropped to maize. Soil Tillage Res. 6:149-161.

Lal, R. 1986. Soil surface management in the tropics for intensive land use and high and sustained production. p. 1-109. *In* Advances in soil science. Vol. 5. Springer-Verlag New York, New York.

Lal, R. 1987. Tropical ecology and physical edaphology. John Wiley and Sons, New York.

Lal, R., and D.J. Cummings. 1979. Clearing a tropical forest. I. Effects on soil and microclimate. Field Crops Res. 2:91-107.

Makarim, A.K., D.K. Cassel, and M.K. Wade. 1988. Effects of land reclamation practices on physical properties of an Oxisol. Soil Tech. 1:195-207.

Maurya, P.R., and R. Lal. 1979. Effects of bulk density and soil moisture on radicle elongation of some soil physical properties and crop production in the tropics. p. 339-347. *In* R. Lal and D.J. Greenland (ed.) Soil physical properties and crop production in the tropics. John Wiley and Sons, New York.

Maurya, P.R., and R. Lal. 1981. Effects of different mulch materials on soil properties and on root growth and yield of maize and cowpea. Field Crops Res. 4:33-45.

McNeil, M. 1964. Lateritic soils. Sci. Am. 211(5):96-102.

Moorman, F.R., and A. Van Wambeke. 1979. Optimum soil utilization systems under differing climatic restraints: Rainy tropical regions. Int. Soil Sci. Soc., 11th Congr. 2:271-291.

Mullins, C.E., D.A. MacLeod, K.H. Northcote, J.M. Tisdall, and I.M. Young. 1990. Hardsetting soils: Behavior, occurrence, and management. *In* Advances in soil science. Vol. II. Springer-Verlag New York, New York.

Nicou, R., and J.L. Chopart. 1979. Root growth and development in sandy and sandy clay soils of Senegal. p. 375-384. *In* R. Lal and D.J. Greenland (ed.) Soil physical properties and crop production in the tropics. John Wiley and Sons, New York.

Norse, D. 1979. Natural resources development strategies and the world food problem. p. 12–51. *In* M.R. Biswas and A.K. Biswas (ed.) Food, climate and man. Wiley Interscience, New York.

Obeng, H.B. 1978. Soil water management and mechanization. Afr. J. Agric. Sci. 5:71–83.

Roose, E.J. 1977. Application of the USLE of the Wischmeier and Smith in West Africa. p. 177–188. *In* D.J. Greenland and R. Lal (ed.) Soil conservation and management in the humid tropics. John Wiley and Sons, New York.

Sanchez, P.A. 1976. Properties and management of soils in the tropics. p. 52–54. John Wiley and Sons, New York.

Sanchez, P.A., W. Couto, and S.W. Buol. 1982. The fertility capability soil classification system interpretation, applicability and modification. Geoderma 27:283–309.

Sanchez, P.A., and J.G. Salinas. 1981. Low-input technology for managing Oxisols and Ultisols in tropical America. Adv. Agron. 34:279–406.

Seubert, C.E., P.A. Sanchez, and C. Valverde. 1977. Effects of land clearing methods on soil properties of an Ultisol and crop performance in the Amazon jungle of Peru. Trop. Agric. 54:307–321.

Silva, L.F. da. 1981. Edaphic changes in "tabuleiro" soils (Haplorthox) as affected by clearing, burning, and management systems. Rev. Theobroma 11:5–19.

Sombreck, W.G. 1984. Soils of the Amazon region. p. 521–533. *In* W. Sioli (ed.) The Amazon: Limnology and landscape ecology of a mighty tropical river and its basin. Dr. W. Junk Publ., Dordrecht, Netherlands.

Stoner, E.R., and E. Freitas. 1989. Characterization of root-restricting zones in Cerrado soils. p. 320–321. *In* Neil Caudle (ed.) Tropsoils technical report, 1986-1987. North Carolina State Univ., Raleigh.

Taylor, H.M. 1974. Root behavior as affected by soil structure and strength. p. 271–291. *In* E.W. Carson (ed.) The plant root and its environment. Univ. of Virginia Press, Charlottesville.

Trangmar, B.B., R.S. Yost, M. Sudjadi, M. Soekardi, and G. Uehara. 1984. Regional variation of selected topsoil properties in Sitiung, West Sumatra, Indonesia. Tropsoils Tech. Rep. 1. Univ. of Hawaii, Honolulu.

Trouse, A.C., Jr. 1979. Soil physical characteristics of root growth. p. 319–325. *In* R. Lal and D.J. Greenland (ed.) Soil physical properties and crop production in the tropics. Wiley-Interscience, New York.

Vine, P.N., R. Lal, and D. Payne. 1981. The influence of sands and gravels on root growth of maize seedings. Soil Sci. 131:124–129.

Williams, B.G., D.J. Greenland, and J.P. Quirk. 1967. The effect of poly (vinylalcohol) on the nitrogen surface area and pore structure of soils. Aust. J. Soil Res. 5:77–83.

Wolf, J.M., and M. Drosdoff. 1976a. Soil water studies in Oxisols and Ultisols of Puerto Rico. I. Water movement. J. Agric. Univ. P.R. 60:375–385.

Wolf, J.M., and M. Drosdoff. 1976b. Soil water studies in Oxisols and Ultisols of Puerto Rico. II. Moisture retention and availability. J. Agric. Univ. P.R. 60:375–385.

6 Relation between Climate and Soil Productivity in the Tropics

M. V. K. Sivakumar

ICRISAT
Niamey, Niger

A. Manu

Texas A&M University
College Station, Texas

S. M. Virmani

ICRISAT
Patancheru, India

E. T. Kanemasu

University of Georgia
Georgia Station, Georgia

There is a heightened concern about producing sufficient food for the current increase in world population. Two alternatives for boosting food production are to increase yield per unit area or expand agriculture into new areas. On a short-term basis, the second alternative seems more attractive. Wortman and Cummings (1978) quote Food and Agriculture Organization (FAO) data that indicate the global availability of 1.6 billion ha of potentially arable land, most of which is in the tropics.

Divergent statements about tropical climates are found in the literature because of the lack of a clear definition of the characteristics and variability of these climates in relation to agricultural applications. Many misconceptions stem from the fact that the tropical climates range from extreme deserts to evergreen forests. Soil is the medium in which the plant expresses its response to climate at a given location. Climate is one of the active factors and generally the dominant one, although its effect is modified by other factors to give rise to a variety of individual soils in any given climatic region (Brengle, 1982). Essentially, climate and soils are linked together inextricably to determine what can be potentially exploited in a given region.

Copyright © 1992 Soil Science Society of America and American Society of Agronomy, 677 S. Segoe Rd., Madison, WI 53711, USA. *Myths and Science of Soils of the Tropics.* SSSA Special Publication no. 29.

Numerous examples abound in literature that show clearly different cropping possibilities under the same climatic regime because of differences in the soil types under consideration. Similarly, the potential of the same soil type is expressed differently under different climatic regimes. Given the extent of coverage of tropics, it is beyond the scope of this chapter to give an exhaustive treatment of the subject. We have drawn upon several examples to illustrate different points.

CHARACTERISTICS OF THE TROPICS

One of the basic myths arises in the definition of the tropical region itself. In the simplest system of climatic classification, many elements of which were derived from the ancient Greeks, the tropics are regions bounded by the astronomically significant 23°27' North and South parallels. This measurement defines only the extreme seasonal limits of the sun's vertical rays and does not consider rainfall. Hence, no modern climatic classification system accepts this as the only criterion. According to Miller (1971), "Supan delimited the tropics by the mean annual isotherm of 68°F (20°C); actually 70°F (21°C) may be a better limit. Koppen (1931), in defining the tropical belt as having 12 mo above 68°F, accepted the temperature of the coldest months as the boundary criterion. Later he preferred the isotherm of 64°F (18°C), which runs close to mean annual isotherm of 70°F.

Another major myth of the tropics is that crops can be grown throughout the year since temperatures are seldom too low for plant growth. While it is true that crops requiring considerable heat to mature and ripen can be grown at any time of the year, this is possible *only if* there is sufficient moisture. Thus, rainfall variability can be a major factor determining the vegetation types. Based on rainfall, the tropics have been divided mainly into three zones: arid, semiarid, and humid. The limits used for the exact delineation of these zones has been a subject that has attracted much attention. Therefore, several climatic classifications exist in literature, based on precipitation (Koppen, 1936), precipitation and temperature (de Martonne, 1926; Emberger, 1955), precipitation and evaporation (Thornthwaite, 1948; Troll, 1965), and moisture availability period (Hargreaves, 1971). For our purpose, however, we have chosen to treat the subject of climate and soil productivity with reference to arid, semiarid, and humid tropics.

CHARACTERISTICS OF CLIMATE RELATED TO SOIL PRODUCTIVITY

Several climatic factors are related to soil productivity. It is possible to show correlations with rainfall for several different soil characteristics.

Rainfall

In temperate regions, temperature is a major factor in cropping. In the tropics, however, rainfall is the major climatic factor affecting soil productivity. It determines the state of vegetation and the surface water as well as groundwater. It has been established that a definite relationship exists between rainfall and the physical and chemical characteristics of the soil.

Rainfall Pattern

One general myth of the tropics is that rainfall here is generally more abundant, and hence less limiting, than in the temperate regions. One argument in support of this assertion is that annual precipitation in South America (which contains large regions with tropical humid climates) is 1350 mm while North America has an annual precipitation of only 670 mm (Trewartha, 1968). While in absolute terms this is true, the rainfall patterns in tropics are more variable and undependable. Variability is a major concern, but it is primarily the deviations that adversely affect the agricultural production. In regions such as the Sahel, the magnitude and extent of the rainfall deviations is very large. As Bowden (1979) described, no growing season is or will be nearly the same in precipitation amount, kind, or range as the previous growing season. Rainfall is best characterized by its irregularity as can be seen from the annual rainfall variation at Filingue, Niger for the past 84 yr (Fig. 6–1a) which varies from 284 (1984) to 980 mm (1912). In the Sahelian and Sudanian climatic zones of West Africa, below-normal rainfall could persist for 10 to 20 yr as can be seen from 1969 to 1988 in the case of Filingue.

Coefficient of variation (CV) of annual rainfall in the arid and semiarid tropics is 20 to 30% while in the humid tropics it is usually about 10 to 20%. In the humid tropics, negative departures in annual rainfall from the mean still leave actual totals that are in most cases sufficient to support reasonable crop growth and yield (Fig. 6–1b). It is the large positive departures that create conditions of excess moisture detrimental to the crops in the humid tropics.

There is a myth about the humid tropics that water is *never* a limiting factor in this region. Although humid tropics show lower CV for annual rainfall than the semiarid tropics, rainfall in some months is usually less variable in the semiarid regions than in the humid regions. Lawson and Sivakumar (1989) showed that CV for July and August rainfall is much less at the semiarid Ouagadougou, Burkina Faso than in Ibadan, Nigeria located in the humid tropics.

One of the less commonly understood features of rainfall in the arid and semiarid regions is the high spatial variability. Spatial variations in rainfall are easily discernible in the field from the uneven distribution of plant cover. In the tropics, especially in Africa, raingauge placement is so widely dispersed that it makes it difficult to assess meaningfully the spatial variations of rainfall on a regional scale. However, microscale studies in the tropics clearly highlight the problem of spatial variability. On a 2.5-km^2 area at the

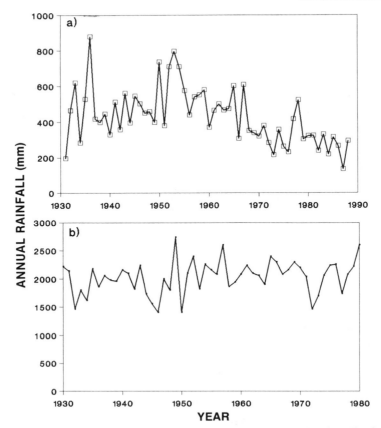

Fig. 6-1. Annual rainfall variation at (a) Filingue, Niger and (b) Benin City, Nigeria.

ICRISAT Sahelian Center, Sadore, Niger, the spatial variability of rainfall was studied using 18 raingages and the extremely high variability in individual storms is shown in Fig. 6-2. Sivakumar and Hatfield (1989) showed that even for seasonal total rainfall, spatial variability persists. This could partially explain the observed variability in plant stands for crops planted on the same day in different fields on the same soil type.

In the tropics, two predominant rainfall patterns, monomodal and bimodal, are observed. In the Southern Sahelian and Sudanian climatic zones of West Africa, rainfall patterns are pronouncedly monomodal. In regions closer to the equator, bimodal rainfall patterns are more common, as in Kenya and parts of Southern India. It is interesting to note that the CV of rainfall in July and August is much higher in the humid bimodal region of Nigeria than in the monomodal rainfall regions of Niger and Burkina Faso (Lawson & Sivakumar, 1989). This could lead a myth that monomodal rainfall patterns are perhaps more desirable than the bimodal patterns since drought spells are less likely under the monomodal regimes because of concentration of rainfall in a single rainy season. Even under the monomomdal rainfall

CLIMATE & SOIL PRODUCTIVITY IN THE TROPICS

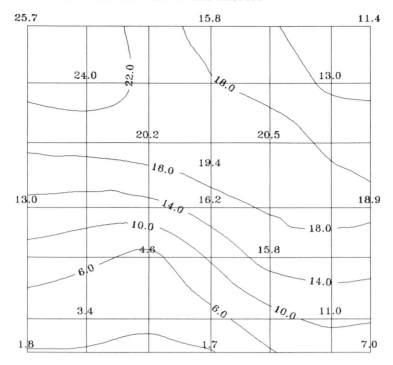

Fig. 6-2. Spatial variability of rainfall on 24 Aug. 1989 at ICRISAT Sahelian Center, Sadore, Niger.

regime, the onset of rains is quite variable as shown in a recent analysis of daily rainfall data in the semiarid regions of West Africa (Sivakumar, 1988). For example, at Niamey, Niger the onset of rains is quite variable and the standard deviation of the onset of rains is 17 d.

Since the ending of rains in the Sahelian and Sudanian climatic zones is sharp and less variable, rainy period is usually shorter in years with delayed onset of rains. In view of the short growing season and the farmer's limited capacity in terms of available power, the number of days available prior to the optimum date of sowing is an important issue. On the sandy soils of the Sahel, evaporation rates are generally high immediately following a rain (Wallace et al., 1988). On such soils, the efficiency with which sowing can be completed before the water evaporates is a critical factor. Hence, the productivity of the soils in this region is significantly influenced by these rainfall events.

Rainfall Intensities and Erosion

Rainfall erosivity is much higher in the tropics than in the temperate regions. Calculations of rainfall erosivity for West Africa by Roose (1977) show that rainfall erosivity factor for the Ivory Coast ranged from 500 to 1400 in contrast to 150 to 650 reported from USA (Wischmeier & Smith, 1960). While only 5 to 10% of the temperate storms are erosive, most of the tropical storms are considered erosive because of the high intensity (Lal, 1977). For example, 82 mm or one-seventh of the seasonal normal rainfall was received in just under 3 h on 4 Aug. 1985 at the ICRISAT Sahelian Center (ISC).

The energy supplied by a storm is determined by the intensity and duration of rainfall. In northern Nigeria, individual rainstorms of >50 mm with peak intensities of 120 to 160 mm/h are not uncommon (Kowal & Kassam, 1977) and peak intensities of more than 250 mm/h for short periods were reported (Kowal, 1970). Peak values of up to 300 mm/h at Niono, Mali (Hoogmoed, 1981) and 386 mm/h at Niamey, Niger (Hoogmoed, 1986) have also been observed.

In the humid tropics, a wide range of rainfall intensities are reported, but higher intensities are mainly associated with thunderstorms (Lal et al., 1980). At Morningside, Tanzania, average rainfall intensity for all months during the year exceeded 20 mm/h while in the rainiest month of April, average intensity exceeded 80 mm/h (Constantinesco, 1976). Peak values in excess of 150 mm/h sustained over 30 min have been observed at Ibadan, Nigeria.

Raindrop impact is a significant factor influencing soil erosion and formation of surface soil crust. During an intense rainstorm, large rain drops break soil aggregates, disperse fine soil particles, and fill surface voids thus effectively sealing the soil surface and reducing the water intake rate (Cannell & Weeks, 1979). The absence of factors able to recreate an aggregated structure, as for instance freezing and thawing do under many temperate soils, means that the deterioration is more or less permanent, unless the soil is put under fallow vegetation for an extended period (Greenland, 1977).

Studies on drop size distribution in Samaru, Nigeria (Kowal & Kassam, 1977) showed that in contrast to the temperate and subtropical regions, brief rainstorms in the tropics are composed of much larger drops. Perrier (1987) reported that during a typical convective storm in the Sahel, wind-driven raindrops striking a dry bare soil surface under the force of a squall or storm front can have an intensity of about 1000 mm/h for a short period where raindrops are driven downward at a velocity of about 60 cm/s. Estimates by Kowal and Kassam (1977) from Nigeria suggest an average kinetic energy load of 34.6 J/(m^2 mm^1). Calculations of Finkel (1986a) showed that the falling rain drops generate 576 times as much power as the equivalent amount of water flowing over land. This force is absorbed by the soil surface and falling raindrops break down and disperse the soil aggregates; cause compaction and sealing of soil surface. Medium-textured soils whose silt content is high are most subject to dispersion in water and form the thickest crusts (Brengle, 1982).

Falling raindrops decrease the infiltration capacity of the topsoil in two ways: (i) by compressing later as they strike the soil surface and plugging the soil pores with loosened soil particles, and (ii) by seepage downwards into the soil of moving soil grains (WMO, 1983). Rates of infiltration are affected by soil types, especially when there are problems of soil crusting. On the bare, weakly crusted soil surface of the sandy soils at ISC, infiltration rates of up to 100 mm/h have been reported (ICRISAT, 1984). However, clogging of soil pores by crusting can reduce the infiltration rate by 2000 times the initial infiltration rate (McIntyre, 1958; Wischmeier & Smith, 1958).

Under conditions of high rainfall intensities, runoff and soil loss are quite common. Data compiled from different studies in West Africa (Table 6-1) show that runoff and soil loss vary with location. Cropped soils, as one would expect, showed much lower rates of runoff.

During the process of runoff, raindrops falling on the layer of water flowing down-slope cause turbulence in this water layer, which assists the soil particles to remain suspended. Even gravel with a diameter of up to 1 cm can be transported by such a turbulent water layer (WMO, 1983).

In areas situated in the low-lying plains that fan out from a range of mountains, runoff from the heavy seasonal rainfall in the mountains causes flash floods. These floods cause extensive damage to soils and crops, while erosion on the mountains and the fan slopes causes deep gullies (Constantinesco, 1976).

Erosion studies conducted at Samaru by Kowal (1970) and Kowal and Stockinger (1973) showed that serious erosion hazards result in a decline in soil fertility and cause problems of soil and water management that are largely unknown in the temperate and subtropical climates. At Niangoloko, Burkina Faso, it has been observed that the increase in water erosion from 1.4 t/(ha yr) to 13 t/(ha yr) decreased the yield of millet from 729 kg/ha to 352 kg/ha (FAO, 1977b).

In the humid tropics, clearing the forest for cultivation results in a larger stream flow. At Mbeye in Tanzania, long-term average increases in stream flow of the order of 50% have been recorded compared with the forested control (Edwards & Blackie, 1975). In Kenya, a 51-mm storm in 30 min caused a loss of nearly 3500 m^3 of soil from a 10-ha coffee (*Coffea arabica* L.) field with a slope of 8°.

An increase in rainfall does not necessarily result in an increase in the erosion. Soil texture and structure both influence the erosion processes of detachment and transport. Clay particles are more difficult to detach than sand due to their attractive charges, but are easier to transport once they are dislodged (Cannell & Weeks, 1979). There are other important intervening factors such as soil erodibility, land form (slope, steepness, and shape), and management systems (Lal, 1980).

Landslides can also be initiated by intense rainstorms in mountain areas and could cause extensive damage.

Potential Evapotranspiration

Potential evapotranspiration (PET) or water demand in the arid and semiarid zones is usually high due to the consistently high air temperatures and radiation load (Jackson, 1977). On the other hand, rainfall is erratic. When rainfall exceeds PET, soil moisture reserves are recharged. When rainfall is less than PE, soil moisture reserves are used. In arid and semiarid areas, PE generally exceeds rainfall. At semiarid locations such as Niamey, Ou-

Table 6–1. Runoff and soil loss data for selected locations.

Country	Location	Mean annual rainfall	Slope	Treatments	Runoff	Soil loss	Reference
		mm	%		%	t/(ha yr)	
Benin	Boukombe	875	3.7	Millet, conventional tillage	11.7	1.3	Verne & Williams, 1965
Niger	Allokoto	452	3.0	Sorghum, cotton	16.3	8.6	Roose & Bertrand, 1971
Nigeria	Samaru	1062	0.3	Bare soil	25.2	3.8	Kowal, 1970
Nigeria	Ibadan	1197	15	Bare soil	41.9	229.2	Lal, 1976a
				Maize-maize	13.5	40.7	
				Maize-cowpea	2.6	0.1	
				Cowpea-maize	12.7	43.0	
Senegal	Sefa	1300	1.2	Bare soil	39.5	21.0	Charreau & Nicou, 1971
	Sefa	1241	1.2	Groundnut	22.8	6.9	Charreau, 1972
		1113		Sorghum	34.1	8.4	
		1279		Maize	30.9	10.3	
Burkina Faso	Ouagadougou	850	0.5	Bare soil	40–60	10–20	Roose & Birot, 1970
				Crop	2–32	0.6–0.8	Charreau & Seguy, 1969
				Forest	2.5	0.1	Charreau & Nicou, 1971
Ivory Coast	Bouake	1200	0.3	Bare soil	15–30	18–30	Charreau, 1972
	Abidjan	2100	7.0	Bare soil	38	108–170	Charreau, 1972
Mali	Niono		1–3	Bare soil	25		Hoogmoed & Stroosnijder, 1980
Niger	Sadore	560		Millet	1.5		Klaij & Serafini, 1988
				Bare soil	0–20		

agadougou, and Hyderabad, rainfall exceeds PE for only 2 to 3 mo (Fig. 6-3). Ibadan, Nigeria, although in humid tropics, shows a 1-mo period in the rainy season when rainfall is less than PE (Fig. 6-3).

The classical work of Denmead and Shaw (1962) showed that the relationship between soil moisture content and transpiration varied with the evaporative demand. On a clear, dry day when the potential evaporation was as high as 6 to 7 mm/d, the decline in transpiration occurred at a very low soil moisture tension. On a heavily overcast day with an evaporative demand of 1 to 4 mm/d, transpiration rate did not decline until the soil moisture reached 120 J kg^{-1}, not much above the permanent wilting point. The erratic rainfall distribution and unpredictable droughts in the arid and semi-arid tropics create conditions of high evaporation demand even in the middle of the rainy season and crops could be exposed to considerable stress.

Relationship Between Rainfall and Soil Moisture Storage

One of the myths of tropics is that since rainfall is abundant, practically any crop can be grown. The precipitation actually stored in the soil depends upon the soil type, surface conditions, and moisture status. Holmes (1961) showed that moisture depletion from a given soil varies with the soil type. The soil profile serves as means of balancing, over time, the discontinuous water supply with a continuous atmospheric evaporative demand (Russell, 1980), which eventually affects plant growth. Because of the great diversity in texture, structure, type of clay, organic matter content and depth of tropical soils, the available water in the soil varies greatly. Weekly soil moisture storage calculations for four soil types of Hyderabad region in India (Fig. 6-4) demonstrate this clearly. In shallow Alfisols, there is little soil moisture storage over extended drought periods. In deep Alfisols, medium deep Vertisols, and in deep Vertisols, there is a fair degree of moisture storage for a substantially longer time during the growing season compared with the shallow Alfisols. Thus, under identical rainfall conditions, the effects of short-term intraseasonal droughts on crop moisture status will differ on different soil types. Therefore, rainfall alone is not an adequate guide to agricultural potential. It is necessary to work out the soil water balance and select crops that grow and mature in the short periods of favorable soil water regime.

Length of the Growing Season

From the above discussion, it should be apparent that the length of growing season, which is a balance between the water supply and demand, varies depending upon the rainfall and the soil type, particularly its moisture-holding characteristics. Calculations of the length of the growing season for three soil types at Hyderabad (Table 6-2) show that on shallow Alfisols with a low moisture-holding capacity, the available growing season is much shorter than on deep Vertisols. This would mean that long duration crops have a higher probability of success on the deep Vertisols than on the shallow Alfisols.

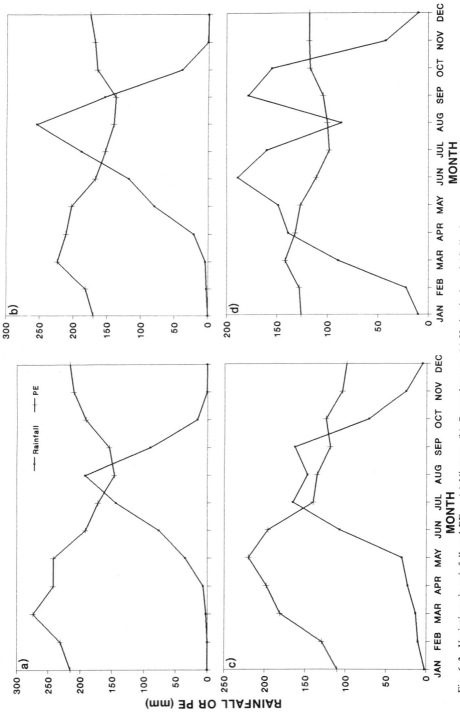

Fig. 6-3. Variations in rainfall and PE at (a) Niamey, (b) Ouagadougou, (c) Hyderabad, and (d) Ibadan.

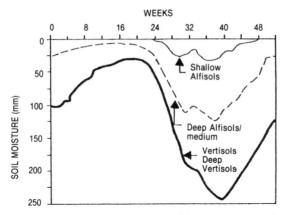

Fig. 6-4. Weekly soil-moisture storage in three soils (based on Hyderabad data, 1901–1970). Source: Virmani (1980).

Drought and Soil Moisture Availability

Droughts in the arid and semiarid tropics are set off by rainfall deviations that fall far below the average rainfall which is already low and undependable. Length of the dry spells that are likely to occur during the growing season can be calculated from historical rainfall data (Sivakumar, 1991). In the Southern Sahelian and Sudanian climatic zones of West Africa, dry spells are most likely to occur in the early and later parts of the growing season. The soil type and its water-holding characteristics play an important role here since droughts early in the growing season occur after the long dry season. The capacity of the soil to retain water, even from small rain showers, thus become critical. On the light, sandy soils that show significantly high evaporation rates immediately after a rain (Wallace et al., 1988), these early season droughts can lead to crop mortality and force the farmers to resow their crops.

When droughts occur during the grain-filling stage, antecedent soil moisture condition is critical to sustain crop growth and productivity. Hence on soils with a higher water-holding capacity, crops can withstand the drought for a much longer period.

Table 6-2. Length of the growing season (in weeks)† for three soil conditions.‡ Source: Virmani (1980).

Rainfall probability	Available water-storage capacity		
	Low 50 mm	Medium 150 mm	High 300 mm
Mean	18	21	26
75%	15	19	23
25%	20	24	30

† From seed-germinating rains (25 June) to the end of season (time when profile moisture reduces AE/PE to 0.5).
‡ Low: shallow Alfisol; Medium: shallow to medium-deep Vertisols; High: deep Vertisol.

In the bimodal rainfall areas of semiarid regions of India, dry spells of varying lengths impose a restriction on the crop choice. In regions where the intervening dry spell is longer than 1 mo, two distinct rainy periods are recognized (southwest monsoon from June to September and northeast monsoon in October to November) but exploitation of these periods depends largely on the soil type. On deep Vertisols with higher water-holding capacity, a long duration legume crop is intercropped with a short duration cereal so that the cereal crop is harvested in the main southwest monsoon season and the legume continues through the shorter northeast monsoon season. The shallow Vertisols are generally left fallow in the southwest monsoon season and are cropped on residual moisture in the northeast monsoon season.

In the Savannah region of West Africa, long dry spells result in sparse vegetation. When this vegetation is completely removed through uncontrolled grazing, and slashing and burning, it leads to a hardened soil surface. Consequently, it sets off the process of long-time soil degradation.

Waterlogging and Leaching

When water in the soil surface due to rainfall, runoff, or irrigation is in excess of the rate of infiltration, and when the topography is enclosed or very flat, temporary waterlogging of the surface or leaching (if the soil horizons are permeable) could occur. Although factors such as topography, physical characteristics of the soil, and influence of soil use by people play an important role in promoting waterlogging, alternating seasons of wet and dry periods could create conditions conducive to waterlogging. In the humid tropics, leaching of clay colloid fractions consisting of clay mineral and Fe and Al hydroxides from the upper horizons leads to acidification. Percolation of these fractions through the profile produces morphological and physicochemical changes. When waterlogging and leaching are particularly pronounced, imbalances may occur in the functioning of the soil and these may sometimes be big enough to be termed degradation (FAO, 1977b).

The evolution of soils such as gleysols, planosols, histosols, and partly vertisols is influenced by waterlogging.

Salinization and Alkalinization

The level of groundwater and its chemical composition are essential factors influencing salinization and soil degradation (WMO, 1983). Groundwater level at a particular location is linked to local climatic conditions. In the arid regions where rainfall is much less than evaporation, salts in solution are drawn upwards by capillary action leading to the formation of saline soils. Also, the salt spreads after the humid years (Kelley, 1963). Weathering of rocks containing sodic minerals causes accumulation of salts on the spot or is transported into low-lying areas. Extensive saline soils around the Lake Chad area is a good example (Cheverry, 1974).

In the semiarid and subhumid tropics, clay leaching and accumulation further down the profile results in the formation of solonetz that quickly causes the development of hydromorphy (WMO, 1983).

Temperature

The importance of temperature in influencing soil productivity in the tropics can be related to von't Hoff's temperature rule that states that for every 10 °C rise in temperature, the velocity of chemical reaction increases by a factor of two to three. The deep-weathered mantle of the tropical regions could be related to the extreme physical and chemical weathering of parent material promoted by high rainfall and temperature. This reflects in the relatively high clay contents, increased aggregation and favorable soil structure in the humid tropics as compared to the sandy structure, poor aggregation, and structure associated with the semiarid soils.

There is an intimate relationship between rainfall incidence and air temperature patterns in the tropics. There is a myth that temperature is not limiting to crop growth in the tropics. However, high temperatures can have a detrimental effect on crop growth in many tropical regions. In the arid and semiarid regions where the onset of the rainy season is preceded by a long dry season, environmental conditions during the stage of crop establishment are usually harsh. From the analysis of frequency distribution of air temperatures in West Africa, Sivakumar (1987) showed that mean maximum temperatures could exceed 40 °C at the time of sowing and that absolute temperatures could be much higher. Although one or two showers can fulfill the moisture requirement for sowing, on the sandy soils of the Sahel, the soil moisture evaporates quickly and if a period of dry, clear weather follows, the soil surface temperatures rise rapidly up to 55 °C. Under these conditions, seedling death is quite common leading to plant stands much below the recommended 10 000 hills/ha.

On soils prone to crusting, such as the Alfisols, crust formation is hastened by high soil temperatures and the additional heat of wetting when raindrops strike a dry soil surface (Perrier, 1987). As the soil gets wetter, the thermal conductivity of the sandy soils of the Sahel was shown to have increased to as much as 0.005 cal/(cm s) per °C.

Even in the humid tropics, seedling growth of soybean [*Glycine max* (L.) Merr.], cowpea [*Vigna unguiculata* (L.)], and pigeonpea [*Cajunus cajan* (L.)] has been shown to be adversely affected by soil temperatures exceeding 34 °C; threshold value for maize (*Zea mays* L.) was 30 °C (Lal, 1974).

In the humid tropics, rapid deforestation for human settlement and cultivation is posing problems. In the southern states of Nigeria, Lowe (1974) reported that 8000 ha were cleared annually. This figure was expected to rise to 60 000 ha by 1984. Dense vegetation removed for human settlement and cultivation in the humid tropics leaves the soil exposed to solar radiation causing the soil temperature to rise. Under conditions of high soil temperatures, the organic matter in the soil decreases rapidly through decomposition. It also limits soil microfauna and microbial activity (Jenkinson & Ayanaba, 1977; Lal, 1987).

Wind

Wind as a climatic factor has a significant effect on the soil productivity in the tropics. A popular myth is to associate damaging effects of wind with tropical cyclones, that are common in the coastal areas. Strong winds are capable of causing extensive mechanical injury and desiccation anywhere.

Particularly dangerous to the environment is blowing dust accompanied by the erosion of the topsoil. Wind erosion is catastrophic in the semiarid tropics where a shift in the ecological balance has led to creation of conditions that accentuate wind erosion. Acute food shortages resulting from increased population pressures and drought have led to the use of marginal lands, overgrazing, and removal of trees and shrubs. Using the desertification map of the world, Balba (1980) estimated that about 577 million ha in the world were vulnerable to desertification due to sand movement. The area of very high desertification due to sand movement was about 44 million ha, 15 million ha in the arid zone, and 29 million ha in the semiarid zone.

Wind erosion is not a problem if a continual wet soil surface or lack of strong winds prevails (WMO, 1983). This is more frequent in the arid, treeless regions with sudden fluctuations in weather and an active wind regime, leading to degradation of the soil cover. Soils of most semiarid lands developed under grass on relatively flat topography. The protective grass cover was removed for cropping and under conditions of severe drought, the soil became exposed to wind erosion. In the Sahelian zone, rains are preceded by violent winds that can remove 10 cm or more of unprotected soil from the field in a few hours that leads to the exposure of highly acidic and sometimes impermeable subsoil.

The main feature of the wind regimes in the Sudanian and Sahelian climatic zones of West Africa is the distinction between the dry and wet seasons (Davy et al., 1976). During the dry season, the harmattan winds blow from the desert areas northeast of the region. In the rainy season, the monsoon regime brings humid winds from the Atlantic Ocean and equatorial Africa to the Southwest.

Surface wind speed and its duration are major factors in the formation of dust storms and the magnitude of dust flow increases with the cube of the wind speed (WMO, 1983). Average wind speeds during the dry season are generally higher, but record wind speeds for the year are expected during thunderstorms early in the rainy season. Wind speeds exceeding 100 km/h have been recorded at ISC. Kowal and Kassam (1978) reported maximum speeds of 110 km/h at Samaru, Nigeria. Wind velocity accompanying a storm influences its kinetic energy and hence the erosive capacity (Lal, 1977). According to Lyles et al. (1969), soil detachment is much higher at higher wind speeds and hence wind-driven rain is more effective in clod breaking. In the Sahelian zone, an enormous amount of dust from the bare, loose, sandy soils is carried in the air because of high wind speeds (Fig. 6-5). This thunderstorm type of dust-raising mechanism is connected with the systematic ascent of air in the convective currents of a storm cloud. A wall of dust several hundred meters high is frequently raised along the thundery squall line. Dur-

Fig. 6-5. Dust storms ahead of rains in the beginning of the growing season are common in the Sahel.

ing rainfall this sand is deposited on the young millet seedlings (Fig. 6-6). The weight of the sand and high soil temperature up to 50 °C in the sand covering the seedling are often fatal to the seedlings (Fig. 6-7) and lead to problems of crop establishment. Downes et al. (1977) established a strong correlation between damage to young vegetable crops and a factor called total kinetic effect (TKE) which is a single expression combining wind velocity, soil flux density, and time of exposure of the seedlings.

The erosive capacity of the wind is modified, among other things, by the erodibility of the soil. This is influenced by soil texture, structure, moisture, and organic matter (Finkel, 1986b). Chepil and Woodruff (1963) found that the erodibility index varies with the amount of clay. Gravelly soils lying in a thin layer over coarse gravel are particularly susceptible to erosion. Sandy and sandy loam soils are also easily prone to erosion. Soil moisture influences the erodibility of soil. Adsorbed moisture forms a film around soil particles and the cohesion between the films creates a resistance to detachment which the force of the wind has to overcome.

Fig. 6-6. Sand covering young millet seedlings after a dust storm.

Wind also affects soil dessication, evaporation being hastened by the removal of water vapor from the soil cavities and the soil surface. Wind action is more effective with a drier atmosphere (WMO, 1983). Wind plays an important role in the extension of saline soils and soil degradation by salinization by trapping the salt spray at the surface of the ocean.

Fig. 6-7. High soil temperatures in the beginning of the growing season are fatal to the young millet seedlings.

Impact of Climatic Factors on Soil Fertility

The implications of the climatic characteristics described earlier in the tropics to soil fertility are both direct and indirect. Increase of organic matter content with rainfall has been demonstrated by several researchers including Gavaud (1968) and Kadeba (1970). Using data from the African tropics where mean annual rainfall ranges from 350 to 1900 mm, Jones (1973) obtained the following relationship between organic C content of soil (C) and rainfall (R):

$$\%C = 0.137 + 0.000865 \times R \; (r = 0.514)$$

Gavaud (1968) also indicated that the C/N ratios tend to decline with decreasing rainfall. He obtained remarkably low C/N values at the limits of the semiarid tropics.

High rainfall regimes of the humid areas lead to high rates of leaching bases with the simultaneous appearance of H and Al ions to compensate for the reduction of bases with respect to the saturation capacity. This leaching of bases influences the soil productivity by creating acidic environment that chemically limits the rooting depth of Al-sensitive crops. The Al released during this process can also influence the P supply. Exchangeable Al reacts with P, both native and applied, thus limiting the availability of P to the plants.

Oxisols and Ultisols that form the bulk of the soils of the rain forest zone, although reasonably well structured with good porosity, low bulk density, and high infiltration rate under native forest vegetation, undergo rapid physical degradation when cleared and exposed to the high-intensity rain and traffic (Lal, 1981). Nutrients in these soils are concentrated in the top 20 cm and sheet erosion selectively removes the organic and mineral colloids and nutrient elements (Roose, 1973; Lal, 1976b). The resulting shallowness of rooting depth affects moisture availability and fertilizer-use efficiency.

Sandy soils of the semiarid tropics have high infiltration and permeability rates. These conditions permit the downward percolation of plant nutrients that leads to further impoverishment of the already infertile soils.

In the semiarid tropics, high-intensity rains promote the accelerated removal of shallow fertile topsoil. Eroded material can take one of two courses: (i) they can be totally lost through runoff and (ii) under certain circumstances they can be deposited within short distances. The consequence of the short-range deposition of material leads to spatial variability in the soil physical and chemical properties over short distances. This results in uneven crop growth, a phenomenon that is prevalent in the sandy Sahelian soils, as documented by Wendt (1986) and Pfordresher et al. (1989).

Under conditions of waterlogging in the humid tropics, excess water in the soil induces an evolution towards an anaerobic medium. Microorganisms responsible for biodegradation of organic matter in an aerobic medium are either destroyed or inhibited. In an anaerobic medium, pH becomes acidic due to the freeing of organic acids and dominant chemical reactions are those of reduction.

Although the low organic C content in the soils of the semiarid tropics could be linked to the low production of biomass, the effect of temperature in hastening the decomposition of plant residue can be an extremely important factor in influencing organic matter equilibrium.

There is a myth that wind erosion is always disastrous. It must, however, be mentioned that wind can act as a fertility renewal factor in the semiarid tropics. Studies conducted by the TROPSOILS team in Niger have shown that incoming wind-blown dust from remote sources is richer in clay, plant nutrients, and has a high pH. Vine (1949) also reported introduction of clayey, ferruginous dust materials into sandy soils in parts of West and East Africa.

Effects of Seasonal Changes in Climate on Soil Productivity

In semiarid regions where a strongly alternating climatic regime of wet and dry seasons occurs, the seasonal changes can influence soil productivity. For example, when dry, Vertisols shrink and crack widely so that the topsoil material can fall down these cracks. In the following rainy season, the clays swell upon wetting. Because there is extra material in the lower horizons, pressure is exerted upwards to give a micro-relief on the soil surface. The swelling and shrinkage associated with these soils has implications with regard to residue management, root development, and water management.

IMPROVED UNDERSTANDING OF THE RELATIONSHIP BETWEEN CLIMATE AND SOIL USING CASE STUDIES

Sandy Soils of the Sahel

In the Sahel, rainfall is variable and undependable. The major soils are generally sandy in texture. For example, soils of Niger are very sandy with the sand fraction usually exceeding 92%. The soil reaction is slightly to strongly acidic and exchangeable Al can exceed 50% of the cation exchange capacity (CEC) in some of the soils. Average water-holding capacity varies, depending upon the depth, from 75 to 150 mm. Poor fertility is a major problem on these soils. Organic matter content rarely exceeds 0.3% and CEC generally <3.0 $cmol_c/kg$. Poor structural stability of these soils is a major constraint since they are susceptible to wind erosion when dry. Under traditional farming conditions, average grain yields of millet are very low, ranging from 130 to 285 kg/ha (Spencer & Sivakumar, 1987).

The above conditions have led to a popular myth that Sahel means most unproductive, dry sand. However, research has shown that increasing crop yields under these conditions is based largely on an improved understanding and judicious utilization of the soil and climatic resources.

From the studies on soil physical, chemical, and hydraulic properties carried out by Hoogmoed and Klaij (1989), a clear understanding of the soils at ISC and their limitations emerges as follows:

1. The soil has a very low fertility status and a rapid decline in its productivity could occur under conditions of continuous cropping without organic or inorganic nutrient addition.
2. Infiltration and redistribution of rainfall on these soils is very rapid, and runoff losses are generally low.
3. Mechanical treatment of the soil is effective only when the soil is moist.
4. The high rates of soil evaporation, especially with rains in the beginning of the season, demand rapid land preparation for sowing and efficient methods of planting.

Climatic Data Analysis for Deriving Effective Cropping Strategies

Prediction of Rainy Season Potential. In view of the large rainfall variability, reliance on monthly and annual rainfall totals is of little value in deriving effective cropping strategies. Parameters such as the onset and ending of rains and the length of the growing season are important for decision making. From a study of these parameters, Sivakumar (1988) showed that it is possible to predict the rainy season potential in the Sahelian zone from the date of onset of rains. This is based on the finding that the onset of rains is much more variable than the ending of rains. Therefore, an early onset of rains offers the probability of a longer growing season while delayed onset results in a considerably shorter growing season. Hence, the potential of the growing season can be assessed with reference to the date of onset of rains. This is illustrated in Table 6-3 for Niamey, Niger (database 1904–1984). If the onset of rains occurs 20 d early (i.e., by 24 May) there is a 71% probability that the growing season will exceed 120 d. On the other hand, if rains are delayed until the beginning of July, there is only little probability of the growing season exceeding 100 d.

Analysis of the Nature of Intra-seasonal Droughts. Although the date of onset of rains provides an idea regarding the potential of the growing season, uncertainties still abound as to rainfall distribution within the growing season. The major concerns are when droughts are most likely to occur within the growing season and what the expected length of droughts would be. Long-term daily rainfall data could be analyzed to answer such questions. Sivakumar (1990a) used the specific definition of onset of rains for each year as the sowing date and computed the length of dry spells (or days until the

Table 6-3. Probabilities of growing season length exceeding specified durations for variable onset of rains for Niamey, Niger. Source: Sivakumar et al. (1990).

Date of onset of rains	Length of growing season (d) exceeding:			
	80	100	120	140
24 May	100	98	71	15
2 June	100	91	40	3
12 June	98	71	15	0
22 June	91	40	3	0
2 July	71	15	0	0

next day with rainfall greater than a defined threshold value) and the percentage frequencies of dry-spell lengths. This analysis showed that in the Sahel, dry spells from the stages of emergence to panicle initiation (up to 20 d after sorting) and grain filling of millet (*Pennisetum glaucum* (L.) R. Br.] last longer than those during panicle initiation to flowering (20-60 DAS). The implication of this analysis for soil management is that conservation of soil moisture in the establishment phase of the crop is critical.

Rainfall Analysis for Preparatory Tillage. In view of the short-growing season and the farmer's limited capacity in terms of available power, the number of days available for preparatory tillage prior to the optimum date of sowing is an important issue. Hoogmoed (1986) concluded that the size of rainfall showers relevant for decision making with regard to preparatory tillage is fairly predictable, and one could calculate the total number of days available for preparatory tillage and sowing. At Niamey, Hoogmoed and Kliaj (1989) showed that the total number of workable days is 31 and the average number of plantable days does not exceed 10 d. This analysis shows that the speed with which planting operations can be carried out is an important issue and the use of animal traction is one of the ways to ensure this.

Strategies for Exploitation of Soil and Climatic Resources

Effective Soil Tillage Methods. On the sandy soils at ISC, beneficial effects of cultivation have been mainly attributed to enhanced rooting in time and space (Klaij & Hoogmoed, 1987). In addition, interrow cultivation during the rainy season helps in weed control and at the end of the season kills weeds and can help in saving precious soil moisture for the subsequent rainy season crop.

Ridging and planting millet on the ridges have been shown to improve the seedling survival compared to the conventional flat planting (Kliaj & Hoogmoed, 1987). An additional advantage of this land treatment is that it facilitates interrow cultivation between ridges using animal traction.

Despite the advantages offered by tillage, it is important to recognize that when the soil is dry, tillage may create a loose top layer thus increasing its susceptibility to wind erosion. On the other hand, tillage under moist soil conditions gives more stable aggregates (Hoogmoed & Klaij, 1989).

Use of a Limited Quantity of Fertilizer to Sustain Productivity and Increase the Water-use efficiency (WUE). Lack of P constitutes a major constraint to crop growth in the Sahel (Jones & Wild, 1975). Results of studies conducted at ISC by the International Fertilizer Development Center (IFDC) showed that application of as little as 20 kg of P_2O_5 doubled the millet yields (Bationo et al., 1987). An important consequence of the use of fertilizer is increased WUE (Sivakumar, 1987).

Millet in the Sahel is traditionally grown in wide rows. Recent investigations at ISC, in collaboration with the Institute of Hydrology (Wallace et al., 1988), showed that water losses in such a system through soil evaporation could be a significant component of the total evapotranspiration (ET).

One of the effective means of reducing evaporative losses is through the use of mulching or crop residues. Recent results from ISC (Bationo et al., 1987) showed that crop residues also help reduce the Al and H saturation of the exchange complex, a major problem on the sandy soils.

Use of Improved Crop Strategies. In view of the rainfall variability in the Sahel, early maturing cultivars have a better chance of success in most of the years. Our studies on drought spells showed that millet cultivars that mature in 90 to 95 d are more likely to avoid the long dry spells during the grain-filling phase than the long duration cultivars of millet that mature in 110 d.

Research at ISC showed that intercropping of millet/cowpea is a useful cropping system. By manipulation of one or more agronomic components such as plant density, spacing etc., substantial yield increases in the intercropping systems can be achieved.

In years with early onset of rains, a complementary system to intercropping is the relay cropping of millet with cowpea for hay. In a field test at ISC, Sivakumar (1990b) showed that with early onset of rains, it is possible to establish a second crop of cowpea for hay. Cowpea enables effective use of the September rains, a large part of which would otherwise have been lost through evaporation because of the poor water-holding capacity of the soils at ISC and provides valuable hay (and possibly grain if rains continue into October) to the farmer.

Synergistic Effects of Different Components

Strategies for exploitation of soil and climatic resources discussed above demonstrate their value, but there is early evidence from the integrated operational scale research (OPSCAR) at ISC that a combination of these inputs offers much greater advantages, while sustaining the yields (ICRISAT, 1989). Field studies conducted at ISC from 1986 to 1988 evaluated the relative advantages offered by one or more combinations of fertilizer, improved varieties of millet and cowpea, animal traction for land preparation and weeding, and rotation of millet with cowpea. From the point of sustainable agriculture, the effect of rotation was most striking in comparison to other inputs such as improved variety along with fertilizer and animal traction. While inputs such as use of an improved millet variety and addition of 13 kg of P/ha gave a 78% yield advantage over the traditional practice, the addition of a rotation treatment in the system boosted the yield advantage to 244% for the millet grain and 95% for the millet straw. These data demonstrate the value of effective soil and water management on the poor, sandy soils.

Deep Vertisols of India and Africa

Vertisols are dark-colored clayey soils that are found under varied climatic conditions covering about 310 million ha worldwide (Dudal, 1965). In Asia they cover much of India (70 m ha) and parts of Burma and Thailand. Several countries in Africa and Latin America have Vertisols and soils of vertic properties. In the tropical countries, these soils present serious problems of land and water management for increased productivity. During the dry season, these soils crack—cracks may be 5 to 20 cm in width and 30 to 50 cm in depth. Therefore, preparation of seedbed is difficult until the onset of the rainy season. Once the rainy season sets in, these soils present traffic problems for land preparation. It is difficult to maintain their surface configuration and since their terminal infiltration rates are extremely low, waterlogging during the rainy season is common.

Due to the constraints previously described, a popular myth prevailed in India that Vertisols cannot be cropped in the rainy season and the prevailing practice has been to leave the soils fallow in the rainy season and crop them in the postrainy season. Hence, the productivity of these soils remained generally low, for example, in the Deccan plateau of India average sorghum [*Sorghum bicolor* (L.) Moench] yield is about 700 kg/ha.

Vertisols of the semiarid tropics, however, have a fairly high potential for crop production that remains to be realized. The case study of Vertisols is illustrated primarily from the studies conducted at the ICRISAT Center located in Patancheru, near Hyderabad, India. This case study shows that through an understanding of climate and the application of science and technology of land, water, and crop management, Vertisols can be cropped in the rainy season and can be made productive.

Land and Water Management

Improved land and water management practices are applied for alleviating the constraints, such as waterlogging, which arise due to the physical properties of Vertisols. Vertisols have very poor internal drainage when they are wet. Under the improved system of management, microwatersheds of 3 to 15 ha size were taken as units for land and water management and agronomic practices. Surface drainage is improved through the provision of surface drains and land smoothing. The in-situ water conservation improvements are brought about by laying out the bed-furrow (ridge-furrow) cultivation systems along the contour. Since the surface runoff water is discharged in a controlled manner, the loss of soil is considerably reduced and WUE is increased considerably. At the ICRISAT Center, the main features of this system are that on a slope of 0.4 to 0.6% graded broadbeds and furrows (50-cm apart) are made that lead into grassed waterways and finally into a dug tank or drain. By following this system, the soil moisture storage is increased and the drainage of excess water is facilitated.

Primary tillage to prepare a rough seedbed is best carried out soon after the harvest of the postrainy- or rainy-season crops. Land should be harrowed whenever 20 to 25 mm of rain is received over a period of 1–2 d. When blade harrowing is done, the clods easily shatter and a satisfactory seedbed is attained.

Dry Sowing Ahead of Onset of Rainy Season

Since the preparation of the seedbed and the sowing of crops present serious problems in Vertisols, the planting of crops in dry soils ahead of the commencement of rains was found to ensure early establishment, and avoid the difficultues associated with planting in a wet, sticky soil. Dry seeding was found successful where the early season rainfall is fairly dependable and when seeds are placed at a depth of 7 to 10 cm. At the ICRISAT Center, good stands were established by dry seeding of crops such as green gram [*Vigna radiata* (L.) R. Wilczek], sunflower (*Helianthus annuus* L.), maize, sorghum, and pigeonpea.

Improved Cropping Systems

The adoption of improved cropping systems provides a continuum of crop growth from the commencement of the rainy season until most of the available moisture is used by the crop. At ICRISAT this was achieved by:

1. Intercropping of long-duration crops (e.g., pigeonpea) with short-duration crops (e.g., maize, sorghum, or soybean).
2. Sequential cropping of crops (e.g., sorghum or maize followed by chickpea or safflower).

Fertility Management

In the tropics, the management of soil fertility is important for realizing the full potential of improved cropping systems. At the ICRISAT Center, effective management of soil and fertilizer N was found to be a necessary ingredient for improved productivity of Vertisols. Application of phosphates and Zn was also found to be essential. Inclusion of legumes in the crop rotations or in intercrop systems was found to have substantially reduced the fertilizer N needs (by about 40 kg of N/ha) of the subsequent cereal crops.

Efficient Farm Machinery

For a successful implementation of the improved Vertisols management system, it is necessary to carry out all the operations thoroughly and in good time. Since animal draught is the main source of energy available to small farm operators of semiarid areas in Asia and Africa, much of ICRISAT's work is related to animal-drawn equipment. Use of a wheeled tool carrier (e.g., Tropicultor or Nikart) was found to be an efficient technique for managing Vertisols in India.

Appropriate Crop Management

To realize the full potential of improved land and water management and cropping systems, it is essential that an appropriate set of crop management practices be adopted. Weed control, integrated pest management, the placement of fertilizers at an appropriate depth and their application at crit-

Table 6-4. Grain yields under improved and traditional technologies on deep Vertisols† at ICRISAT Center‡ in 13 successive yr.

		Grain yield, t/ha				
		Improved system: Double cropping			Traditional system, Single crop	
Year	Cropping period rainfall, mm	Sorghum/ Maize	Sequential or chickpea intercropped pigeonpea	Total	Sorghum or chickpea	
1976/77	708	3.20	0.72	3.92	0.44	0.54
1977/78	616	3.08	1.22	4.30	0.38	0.87
1978/79	1089	2.15	1.26	3.41	0.56	0.53
1979/80	715	2.30	1.20	3.50	0.50	0.45
1980/81	715	3.59	0.92	4.51	0.60	0.56
1981/82	1073	3.19	1.05	4.24	0.64	1.05
1982/83	667	3.27	1.10	4.37	0.63	1.24
1983/84	1045	3.05	1.77	4.82	0.84	0.48
1984/85	546	3.36	1.01	4.37	0.69	1.23
1985/86	477	2.70	0.73	3.43	§	0.84
1986/87	585	4.45	0.38	4.83	0.37	1.27
1987/88	841	4.26	1.35	5.61	0.80	0.92
1988/89	907	4.64	1.23	5.87	0.61	1.18
Mean	771	3.33	1.07	4.40	0.59	0.86
SD	205	0.76	0.34	0.76	0.15	0.32
CV, %	27	23	32	17	25	37

† Available water-holding capacity 150 cm per m of soil depth.
‡ Average rainfall for Hyderabad (29 km away from ICRISAT Center) based on 1901-1984 data is 784 mm with a CV of 27%.
§ No crop sown.

ical stages of crop growth are some of the crop management factors that could lead to the realization of high and sustained yields on Vertisols.

One important aspect of the improved Vertisol technology is the synergistic effect of various components when applied together, as compared with their individual effect. This point has been brought out convincingly after 13 yr of watershed-based experimental results from ICRISAT (Table 6-4). Kanwar and Rego (1983) and Kanwar et al. (1982) noted during the steps taken to improve Vertisols technology conducted at ICRISAT that though the contribution of fertilizers was highest, the response to fertilizers was most highly marked when they were applied in combination with improved land and water management treatments and the adoption of improved agronomic practices. This observation has great relevance in the African continent. Here fertilizers are costly and in most instances have to be imported. All efforts, therefore, must be made to realize maximum fertilizer-use efficiency by applying the principles of improved Vertisol technology.

Improved Productivity

Use of the above components of technology made it possible to grow two crops, one in the rainy season and another in the postrainy season and resulted in considerable increases in crop production (Virmani et al., 1989). Where a farmer harvested about one-half ton of sorghum or chickpea by

Table 6-5. Grain yields of some cropping systems on vertic soils† under low (0-0-0) and medium (60-12-0) fertility at ICRISAT Center in operational scale experiments. Source: ICRISAT (1983, 1984).

Year	Cropping period rainfall, mm	Soil fertility	Grain yield, kg/ha				
			Sole pigeonpea	Sorghum/ pigeonpea	Millet/ pigeonpea	Ground-nut/ pigeonpea	Sole sorghum
1981/82	1073	Low	700	937	1201	1387	516
		Medium	868	2175	3581	1423	3234
198/83	667	Low	1041	2219	2190	2214	1170
		Medium	1217	4291	3178	2917	2869

† Available water-holding capacity of 50 cm soil profile is 80 mm.

using his traditional system, a total yield of about 3 t of grain/ha has been consistently harvested through a two-crop combination under the improved Vertisols management system at ICRISAT during 1976-1989 (Table 6-4). Further, in the vertic soils several intercrop combinations (e.g., sorghum-pigeonpea or millet-pigeonpea) have produced yields of 2 to 3 t/ha under medium fertility treatment (60-12-0), as shown in Table 6-5. The introduction of the new system also has resulted in: (i) a considerable reduction in soil erosion; (ii) a much higher in-situ moisture conservation, and therefore in higher rainfall-use efficiency (Table 6-6); and (iii) much more dependable harvests year after year (Table 6-4).

Table 6-6. Annual water balance and soil loss (t/ha) for traditional and improved technologies in Vertisol watersheds, ICRISAT Center, 1976-77 to 1983-84.

Farming systems technology	Water-balance component				
	Annual rainfall	Water used by crops	Water lost as surface runoff	Water lost as bare soil evaporation and deep percolation	Soil loss
	mm				t/ha
Improved system:					
Double cropping on broadbed and furrows	904	602 (67)†	130 (14)	172 (19)	1.5
Traditional system:					
Single crop in postrainy season and cultivation on flat	904	271 (30)	227 (25)	406 (45)	6.4

† Figures in parentheses are amounts of water used or lost expressed as percentage of total rainfall.

CONCLUSIONS

In this chapter, we have attempted to show the important relationships between climate and soil productivity in the tropics. It is apparent that factors such as rainfall pattern, rainfall intensities and erosion, waterlogging, potential ET, length of the growing season, temperature, and wind should be carefully considered in an assessment of soil productivity in the tropics. Several myths concerning the role of these climatic factors exist and we tried to clarify relevant issues using data from current literature.

The case studies from research carried out at the ICRISAT Sahelian Center in Niger and at the ICRISAT Center in India show that productivity of tropical soils can be much improved by the application of improved understanding of climatic constraints to develop effective soil and water-management practices. Sustainability of tropical agriculture in the future is linked to our ability to understand the limits imported by environmental constraints on soil productivity and appropriately manage these soils while ensuring long-term stability.

REFERENCES

Balba, A.M. 1980. Desertification in North Africa. p. 14–25. *In* Desertification and soils policy. Symposia Papers III. 12th Int. Congr. Soil Sci., New Delhi, India.

Bationo, A., C.B. Christianson, and A. Mokwunye. 1987. Soil fertility management of the millet producing sandy soils of Sahelian West Africa: The Niger experience. p. 159–168. *In* Soil, crop and water management systems for rainfed agriculture in the Sudano-Sahelian zone: Proc. of an Int. Workshop. 11–16 Jan. ICRISAT Sahelian Center, Niamey, Niger. ICRISAT, Patancheru, India.

Bowden, L. 1979. Development of present dryland farming systems. p. 45–72. *In* A.E. Hall et al. (ed.) Agriculture in semi-arid environments. Springer-Verlag, Berlin.

Brengle, K.G. 1982. Principles and practices of dryland farming. Colorado Assoc. Univ. Press, Boulder.

Bridges, E.M. 1978. World soils. Cambridge Univ. Press, London.

Cannell, G.H., and L.V. Weeks. 1979. Erosion and its control in semi-arid regions. p. 238–256. *In* A.E. Hall et al. (ed.) Agriculture in semi-arid environments. Springer-Verlag, Berlin.

Charreau, C. 1972. Problemes poses par l'utilization agricole des soils tropicaux par des cultures annuelles. Trop. Soil Res. Symp. May 1972. Int. Inst. Trop. Agric. (IITA), Ibadan, Nigeria.

Charreau, C. 1974. Soils of tropical dry and dry wet climatic areas and their use and management. A series of lectures, Cornell Univ., Ithaca, NY.

Charreau, C., and R. Nicou. 1971. L'amélioration du profil cultural dans les sols sableux et sablo argileux de la zone tropicale séche Ouest Africaine et ses incidences agronomiques. Agric. Trop. 26 (2):209–225, (5):565–631, (9):903–978, 1183–1247.

Charreau, C., and L. Seguy. 1969. Measure de l'erosion et du ruissellement à Sefa en 1968. Agric. Trop. 20:6–7, 600–625.

Chepil, W.S., and N.P. Woodruff. 1963. The physics of wind erosion and its control. Adv. Agron. 15:211.

Cheverry, C. 1974. Contribution a l'étude pédologique des polders du lac Tchad. Mém. ORSTOM, Paris.

Constantinesco, I. 1976. Soil conservation for developing countries. FAO Soils Bull. 30. FAO, Rome.

Davy, E.G., F. Mattei, and S.I. Solomon. 1976. An evaluation of climate and water resources for development of agriculture in the Sudano-Sahelian zone of West Africa. Spec. Environ. Rep. 9. World Meterological Organization (WMO), Geneva, Switzerland.

de Martonne, E. 1926. Une nouvelle fonction climatologique: L'indice d'aridite. La Météorol. 68:449–458.
Denmead, O.T., and R.H. Shaw. 1962. Availability of water to plants as affected by soil moisture content and meteorological conditions. Agron. J. 54:385–390.
Downes, J.D., D.W. Fryrear, R.L. Wilson, and C.M. Sabota. 1977. Influence of wind erosion on growing plants. Trans. ASAE 20:885.
Dudal, R. 1965. Dark clay soils of tropical and subtropical regions. Agric. Development Paper 83. FAO, Rome.
Edwards, K.A., and J.R. Blackie. 1975. Hydrological research in East Africa. E.A. Agric. For. J. (special issue).
Emberger, L. 1955. Une classification biogéographique des climats. Recueil des travaux des laboratoires des botanique, géologie et zoologie de la Faculté des Sciences de l'Université de Montpellier (Ser. Bot. No. 7):3–45.
Food and Agriculture Organization. 1977a. The FAO/UNFPA Expert Consultation on Land Resources for Population of the Future. FAO, Rome.
Food and Agriculture Organization. 1977b. A framework for land evaluation. FAO Soils Bull. 32 and ILRI Publ. 22. FAO, Rome, and ILRI, Wageningen.
Food and Agriculture Organization. 1977c. Assessing soil degradation. Soils Bull. 34. FAO, Rome.
Finkel, H.J. 1986a. Semiarid soil and water conservation. CRC Press, Boca Raton, FL.
Finkel, H.J. 1986b. Wind erosion. p. 109–121. *In* H.J. Finkel (ed.) Semiarid soil and water conservation. CRC Press, Boca Raton, FL.
Forest, F., and B. Lidon. 1984. Influence du regime pluviométrique sur la fluctuation du rendement d'une culture de sorgho intensifiée. p. 247–261. *In* Agrometeorology of sorghum and millet in the semi-arid tropics: Proc. Int. Symp. 15–20 Nov. 1982. ICRISAT Center, Patancheru, India.
Gavaud, M. 1968. Les sols bien draines sur materiaux sableux du Niger. Cah. Orstom, Ser. Pedol. 6:277–307.
Greenland, D.J. 1977. Soil structure and erosion hazard. *In* D.J. Greenland and R. Lal (ed.) Soil conservation and management in the humid tropics. John Wiley and Sons, New York.
Hargreaves, G.H. 1971. Precipitation dependability and potential for agricultural production in Northeast Brazil. Publ. 74-D159. EMBRAPA and Utah State Univ., Logan.
Holmes, R.M. 1961. Estimation of soil moisture content using evaporation data. *In* Proc. Hydrology Symp. 2. Evaporation. Dep. of Northern Affairs and Natural Resources, Water Resources Branch, Toronto, ON.
Hoogmoed, W.B. 1981. Analysis of rainfall in some locations of West Africa and India. *In* E. Rawitz et al. (ed.) Development of criteria and methods for improving the efficiency of soil management and tillage operations with special reference to arid and semi-arid regions. Tillage Lab., Agric. Univ., Wageningen, Netherlands and Dep. of Soils and Water Science, Hebrew Univ., Rehovot, Israel. Appendix 5.
Hoogmoed, W.B. 1986. Analysis of rainfall data relating to the number of days available for tillage and planting in some selected locations in Niger. Rep. 86-4. Agric. Univ., Wageningen, Netherlands and ICRISAT Sahelian Center, Niamey, Niger.
Hoogmoed, W.B., and M.C. Klaij. 1990. Soil management for crop production in the Sahel. I. Soil and climatic parameters. Soil Tillage Res. 16:85–103.
Hogomoed, W.B., and L. Stroosnijder. 1984. Crust formation on sandy soils in the Sahel. I. Rainfall and infiltration. Soil Teillage Res. 4:5–23.
International Crops Research Institute for the Semi-Arid Tropics. 1983, 1984. Annual Rep. 1983, 1984. ICRISAT, Patancheru, India.
International Crops Research Institute for the Semi-Arid Tropics. 1989. Annual report 1988. ICRISAT, Patancheru, India.
Jackson, I.J. 1977. Climate, water and agriculture in the tropics. Longman, London.
Jenkinson, D.S., and A. Ayanaba. 1977. Decomposition of carbon-14 labeled plant material under tropical conditions. J. Soil Sci. Soc. Am. 41:912–915.
Jones, M.J. 1973. The organic matter content of savanna soils of West Africa. J. Soil Sci. 24:42–53.
Jones, M.J., and A. Wild. 1975. Soils of the West African Savanna: The maintenance and improvement of their fertility. Tech. Commun. 55, Commonwealth Agricultural Bureaux, Farnham Royal, UK.
Kadeba, O. 1970. Organic matter and nitrogen status of some soils from the savanna zone of Nigeria. Proc. Inaugural Conf. Forestry Assoc., Nigeria, Ibadan.

Kanwar, J.S., J. Kampen, and S.M. Virmani. 1982. Management of Vertisols for maximising crop production—ICRISAT Experience. p. 94–118. *In* Symposium Papers. II. Trans. Int. Soc. of Soil Sci. 12th. New Delhi, India. 8–16 Feb. Indian Soc. Soil Sci., New Delhi, India.

Kanwar, J.S., and T.J. Rego. 1983. Fertilizer use and watershed management in rainfed areas for increasing crop production. Fert. News 28:33–43.

Kelley, W.P. 1963. Use of saline irrigation water. Soil Sci. 95:385–391.

Klaij, M.C., and W. Hoogmoed. 1987. Crop response to tillage practices in a Sahelian soil. p. 265–276. *In* Workshop on Soil, Crop and Water Management Systems for Rainfed Agriculture in the Sudano-Sahelian Zone. 11–17 Jan. Niamey, Niger.

Klaij, M.C., and P.G. Serafini. 1988. Management options for intensifying millet based crop production systems on sandy soils in the Sahel. p. 501–503. *In* P.W. Unger et al. (ed.) Challenges in dryland agriculture. A global perspective. Proc. Int. Conf. on Dryland Farming. 15–19 Aug. Amarillo, Bushland, TX.

Koppen, W. 1931. Grundriss der Klimakunde. (Second ed. of Die Klimate der Erde). Springer Verlag, Berlin.

Koppen, W. 1936. Das geographische System der Klimate. *In* W. Koppen and R. Geiger (ed.) Handbuch der Klimatologie. Vol. 1, Part C. Gebruder Borntrager, Berlin.

Kowal, J.M. 1970. The hydrology of a small catchment basin at Samaru, Nigeria: Assessment of surface run-off under varied land management and vegetation. Niger. Agric. J. 7:120–133.

Kowal, J.M., and A.H. Kassam. 1977. Energy load and instantaneous intensity of rainstorms at Samaru, Northern Nigeria. *In* D.J. Greenland and R. Lal (ed.) Soil conservation and management in the humid tropics. John Wiley and Sons, New York.

Kowal, J.M., and A.H. Kassam. 1978. Agricultural ecology of savanna: A case study of West Africa. Clarendon Press, Oxford.

Kowal, J.M., and K.S. Stockinger. 1973. The usefulness of ridge cultivation in agriculture. Soil Water Conserv. J. 28:136–137.

Lal, R. 1974. Effect of constant and fluctuating soil temperatures on growth, development and nutrient uptake by maize seedlings. Plant Soil 40:586–606.

Lal, R. 1976a. Soil erosion investigations on an Alfisol in Southern Nigeria. IITA Monogr. I. Int. Inst. for Trop. Agric., Ibadan, Nigeria.

Lal, R. 1976b. Soil erosion on Alfisols in Western Nigeria. I: Effect of slope, rotation and residue management. Geoderma 16:363–375.

Lal, R. 1977. Analysis of factors affecting rainfall erosivity and soil erodibility. p. 49–55. *In* D.J. Greenland and R. Lal (ed.) Soil conservation and management in the humid tropics. John Wiley and Sons, New York.

Lal, R. 1980. Soil erosion as a constraint to crop production. p. 405–423. *In* Priorities for alleviating soil-related constraints to food production in the tropics. Int. Rice Res. Inst., Los Baños, Philippines.

Lal, R. 1981. Soil erosion problems on Alfisols in western Nigeria. VI. Effects of erosion on experimental plots. Geoderma 25:215–230.

Lal, R. 1987. Effects of soil erosion on crop productivity. CRC Crit. Rev. Plant Sci. 5:303–308.

Lal, R., T.L. Lawson, and A.H. Anastase. 1980. Erosivity of tropical rains. p. 143–151. *In* M. Deboodt and D. Gabriels (ed.) Assessment of erosion. John Wiley and Sons, Chichester, England.

Lawson, T.C., and M.V.K. Sivakumar. 1989. Climatic constraints to crop production and fertilizer use. Fert. Res. 29:9–21.

Lowe, R.G. 1974. Shifting cultivation and soil conservation in Africa. FAO Soils Bull. 24. FAO, Rome.

Lyles, L., L.A. Disrud, and N.P. Woodruff. 1969. Effect of soil physical properties, rainfall characteristics and wind velocity on clod disintegration by simulated rainfall. Proc. Soil Sci. Soc. Am. 33:302–306.

McIntyre, D.S. 1958. Permeability measurements of soil crusts formed by raindrop impact. Soil Sci. 85:185–189.

Miller, A.A. 1971. Climatology. Mathuen, London.

Perrier, E.A. 1987. An evaluation of soil-water management on an Alfisol in the semi-arid tropics of Burkina Faso. p. 59–66. *In* Alfisols in the semi-arid tropics. Proc. Consultants Workshop on the State of the Art and Management Alternatives for optimizing productivity of SAT Alfisols and related soils. 1–3 Dec. 1983. ICRISAT, Patancheru, India.

Pfordresher, A., L.P. Wilding, L.R. Hossner, A. Manu, S.C. Geiger, and R.C. Maggio. 1989. Applications of infrared video imagery to evaluate crop spatial variability in the Sahel, West Africa. Dep. of Soil and Crop Sciences/TROPSOILS, Texas A&M Univ., College Station.

Roose, E.J. 1973. Dix-sept années de mesures expérimentales de l'erosion et du ruissellment sur un sol ferrallitique sableux de basse Cote d'Ivoire. Contribution à l'étude de l'erosion hydrique en milieu intertropical. ORSTOM, Abidjan, Thése Doct. Ing., Fac. Sci. Abidjan.

Roose, E.J. 1977. Application of the universal soil loss equation of Wischmeier and Smith in West Africa. p. 177-187. *In* D.J. Greenland and R. Lal (ed.) Soil conservation and management in the humid tropics. John Wiley and Sons, New York.

Roose, E.J., and R. Bertrand. 1971. Contribution a l'etude de la méthode des bandes d'arret pour lutter contre l'erosion hydrique en Afrique de l'ouest. Résultats expérimentaux et observations sur le terrain. Agron. Trop. (Paris) 26:1270-1283.

Roose, E.J., and Y. Birot. 1970. Mesure de l'erosion et du lessionage oblique et vertical sous une savane arbor e du plateau Mossi (Gonse, Haute-Volta Campagnes 1968-69) CIFT.1 ORSTOM. Inst. Francais de Recherche Sci. pour le Developpement en Coop., Paris.

Russell, M.B. 19080. Profile moisture dynamics of soil in Vertisols and Alfisols. p. 75-87. *In* Proc. Int. Workshop on the Agriclimatological Res. Needs of the Semi-Arid Tropics. 22-24 Nov. 1978. ICRISAT, Patancheru, India.

Sivakumar, M.V.K. 1987. Agroclimatic aspects of rainfed agriculture in the Sudano-Sahelian zone. p. 17-38. *In* Workshop on Soil, Crop and Water Management Systems for Rainfed Agriculture in the Sudano-Sahelian Zone. 11-17 Jan. Niamey, Niger.

Sivakumar, M.V.K. 1988. Predicting rainy season potential from the onset of rains in the Sahelian and Sudanian climatic zones of West Africa. Agric. Forest Meteorol. 42:295-305.

Sivakumar, M.V.K. 1990. Exploiting rainy season potential from the onset of rains in the Southern Sahelian zone of West Africa. Agric. Forest Meteorol. 51:321-332.

Sivakumar, M.V.K. 1991. Drought spells and drought frequencies in West Africa. Res. Bull. 13. ICRISAT, Patancheru, India.

Sivakumar, M.V.K., and J.L. Hatfield. 1990. Spatial variability of rainfall on an experimental station in Niger, West Africa. Theo. Appl. Climat. 42:33-39.

Spencer, D.S.C., and M.V.K. Sivakumar. 1987. Pearl millet in African agriculture. p. 19-31. *In* Proc. Int. Pearl Millet Workshop. 7-11 Apr. 1986. ICRISAT, Patancheru, India.

Thornthwaite, C.W. 1948. An approach towards a rational classification of climate. Geogr. Rev. 38:55-94.

Trewartha, G.T. 1968. An introduction to climate. McGraw-Hill Book Co., New York.

Troll, C. 1965. Seasonal climates of the earth. p. 28. *In* E. Rodenwalt and H. Jusatz (ed.) World maps of climatology. Springer Verlag, Berlin.

Verne, R., and P. Williams. 1965. Résultats des études de l'erosion au Dahomey. Communication au Colloque: Conservation et amélioration de la fertilité des sols; Khartoum. Organ. Afr. Unity Sci. Tech. Res. Comm. Publ. 98:43-53.

Vine, H. 1949. Nigerian soils in relation to parent materials. Commonw. Bur. Soil Sci. Tech. Commun. 46:22-29.

Virmani, S.M. 1980. Climatic approach. p. 93-102. *In* Proc. Int. Symp. on Devel. and Transfer of Technology for Rainfed Agriculture and the SAT Farmer. 28 Aug.-1 Sept. 1979. ICRISAT, Patancheru, India.

Virmani, S.M., M.R. Rao, and K.L. Srivastava. 1989. Approaches to the management of Vertisols in the semi-arid tropics: The ICRISAT experience. p. 17-36. *In* Management of Vertisols for improved agricultural production: Proc. an IBSRAM Inaugural Workshop. 18-22 Feb. 1985. ICRISAT Center, Patancheru, India.

Wallace, J.S., J.H.C. Gash, D.D. McNeil, and M.V.K. Sivakumar. 1988. Measurement and prediction of actual evaporation from sparse dryland crops. Scientific Rep. on Phase II of ODA Proj. 149. Inst. Hydrology, Wallingford, UK.

Wendt, J. 1986. Pearl millet response to soil variability in sandy Ustalfs near Niamey, Niger, West Africa. M.Sc. thesis. Texas A&M Univ., College Station.

Wischmeier, W.H., and D.D. Smith. 1958. Rainfall energy and its relationship to soil loss. Am. Geophys. Union 39:285-291.

Wischmeier, W.H., and D.D. Smith. 1960. A universal soil loss equation to guide conservation farm planning. Trnas. Int. Congr. Soil Sci., 7th 1:418-425.

World Meteorological Organization. 1983. Meteorological aspects of certain processes affecting soil degradation—especially erosion. Tech. Note 178. WMO, Genera.

Wortman, S., and R.W. Cummings, Jr. 1978. To feed this world: The challenge and the strategy. Johns Hopkins Univ. Press, Baltimore, MD.

7 Myths and Science of Fertilizer Use in the Tropics

A. Uzo Mokwunye
IFDC-Africa
Lomé, Togo

L. L. Hammond
Texasgulf, Inc.
Raleigh, North Carolina

By definition, tropical environments are those in which the mean monthly temperature, adjusted to sea level, does not dip below 18 °C (Dudal, 1979). Because these mean annual temperatures are ideal for plant growth, tropical environments are supposed to possess enormous potential for agricultural production. The reality, however, is that most countries classified as "developing" are in this belt. It is striking that problems of food production and low income among the rural poor are more commonly found in these regions of the world. Obviously, there is a gap between potential productive power and actual production figures.

One of the reasons for this gap is that inherently poor soil fertility is a major constraint to increased agricultural production. For example, more than 80% of the soils of tropical Africa and America have serious limitations for crop production (Sanchez, 1976). One consequence of the high population growth rates characteristic of these developing countries (Bumb, 1989) is that traditional practices for maintenance of soil productivity are no longer viable. Fertilizers are needed to promote and sustain the ability of the soils to grow crops. The benefits of fertilizers for developing countries, as listed in Fig. 7-1, conclusively demonstrate how indispensable fertilizers are in efforts to increase food production while maintaining the natural resource base (Baanante et al., 1989). However, limited understanding of the complex nature of tropical soils has given rise to several perceptions relative to fertilizer requirements of the crops and soils, fertilizer types, and fertilizer efficiency. It is our intention in this chapter to review some of these myths in light of our present knowledge of the tropical ecosystem.

Copyright © 1992 Soil Science Society of America and American Society of Agronomy, 677 S. Segoe Rd., Madison, WI 53711, USA. *Myths and Science of Soils of the Tropics.* SSSA Special Publication no. 29.

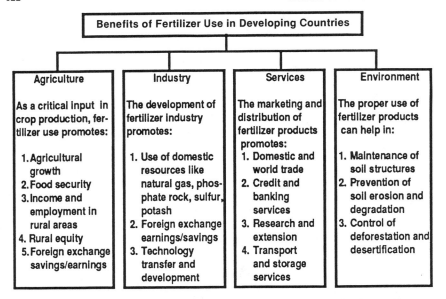

Fig. 7-1. Summary chart of the benefits of fertilizer use in developing countries. Source: Baanante et al. (1989).

THE MYTHS AND SCIENCE OF FERTILIZER USE

Myth or Science? Tropical Soils Have High Phosphorus-Fixation Capacity

As Sanchez (1976) noted, P fixation is an ill-defined term in soil science. In its everyday use, P fixation connotes the retention of available P in a form unavailable to the plant. It erroneously signifies the irreversible transformation of water-soluble monocalcium phosphate, the primary product of commercial superphosphates, into Al, Fe, and Ca phosphates.

Sanchez and Uehara (1980) defined soils with high phosphate-fixing capacity as those that require additions of at least 200 kg of P ha^{-1} to provide an equilibrium soil solution concentration of 0.2 mg of P kg^{-1} of soil. Acid soils that meet this definition are usually those that have loamy or clayey topsoil textures with a sesquioxide/clay ratio of 0.2 or greater or have a dominance of allophane in the clay fraction of the topsoil (Buol et al., 1975). Using these criteria, Cochrane and Sanchez (1981) estimated that about 53% of tropical America's land surface is dominated by soils with high phosphate-fixing capacity. There have been modifications to these criteria. Workers in the Pasture Program at the International Center of Tropical Agriculture (CIAT) have defined soils with high phosphate-fixing capacity as those in which the clay content of the topsoil is >35% and the free Fe oxide/clay ratio is >1.5. A map of the lower central region of tropical South America (Fig. 7-2) clearly indicates that soils with high P-fixing properties do not

FERTILIZER USE IN THE TROPICS

■ High P-Fixing Soils - > 35% clay; % Free Fe_2O_3/ % Clay > 0.15.

□ Soils with Lower P-Fixing Capacity.

Fig. 7-2. Phosphorus-fixing soils of lower central South America. Source: Cochrane and Sanchez (1981).

predominate in this region (Cochrane & Sanchez, 1981). Phosphorus deficiency affects more than 90% of the soils of the Amazon Basin. However, only 16% are subject to high P fixation (Cooke, 1984).

In Africa, Greenland (1973) warned of the undue emphasis put on the phosphate-fixing capacity of the soils of the region. Russel (1974) attempted to discourage the notion that soils of the tropics fixed phosphates in a way different from that of temperate region soils. He indicated that the lower P content of the soils of the tropical climates should not lead to the conclusion that the soils of the tropics reacted any differently from the soils of the temperate regions. Data in Table 7-1 illustrate the P-sorption characteristics of several West African soils. The data indicate that none of the soils can be designated as having high phosphate-fixing capacity. The scientific

Table 7-1. Sorption characteristics of soils from humid, subhumid, and semiarid West Africa. Source: Mokwunye et al., 1986.

Location	Soil	Parent materials	P sorption maximum[†]
			mg kg^{-1}
Jos	Alfisol	Basic ferromagnesium	244
Onne	Ultisol	Acidic crystalline rocks	233
Ikenne	Alfisol	Basement complex rocks	178
Samanko	Alfisol	Intermediate crystalline rocks	172
Kaura Namoda	Alfisol	Intermediate crystalline rocks	112
Sadore	Psamment	Dune sand	36

† Sorption maximum based on Langmuir equation except for Sadore which was actually measured.

reality is that tropical soils belonging to the suborder Andepts, as well as some Oxisols and Ultisols, do fix high amounts of P (Sanchez, 1976). But many of the soils in these orders have coarse-textured surface horizons and do not fix large quantities of phosphate. Most of the Ultisols of the coastal regions of West Africa fall in this category. With the exception of certain Rhodic or Oxic Alfisols and Inceptisols, the soils of subhumid and semiarid tropical Africa have low capacities to retain P.

Myth or Science? Low-Solubility Phosphate Rocks are More Suited for Use in High Phosphorus-Fixing Soils

Because soils of the tropics generally have lower levels of native phosphate and the myth was already established that these soils have a high P-fixing capacity, it was theorized that the ideal way to fertilize these soils with P was to add low-solubility phosphate rocks (PRs). Although there was evidence to show that some PRs are less efficient than soluble P fertilizers even in acid soils, it was argued that the relative agronomic efficiency of the PRs compared with superphosphate would be higher in soils with high P-retention capacity than in those with lower P-retention capacity. The slow PR solubilization would increase the chances for the plant roots to take up the P before it is transformed to unavailable forms (Nunez, 1984).

This myth has been reinforced by experimental evidence showing that the dissolution rate of PRs in soils increases with an increase in P-retention capacity (Smyth & Sanchez, 1982). This same experiment, however, showed that the amount of plant-available P as measured by laboratory extraction procedures (Olsen, Bray 1, and resin) decreased as P-retention capacity increased. Field data reported by Hammond and Leon (1983) indicated that finely ground PRs applied to Latin American Oxisols and Ultisols were relatively more effective with respect to triple superphosphate (TSP) than were the same PR sources applied to Andepts, which exhibited significantly higher P-retention capacities than the Oxisols and the Ultisols (Table 7-2). The data also showed that the Fosbayovar PR from Peru, which is more reactive than the Pesca rock from Colombia, gave the highest relative agronomic effectiveness in both low P-fixing and high P-fixing soils.

Table 7-2. Properties of Colombian soils and the relative agronomic effectiveness (RAE) of P sources as measured by dry matter production of *Brachiaria decumbens*.

	Reactive				RAE†		
	pH	Bray I	Al	P fixation	Fosbayovar PR‡	Pesca PR	Pesca PR
		mg P kg^{-1}	mmol kg^{-1}	% at 100 mg kg^{-1}			
I. Ultisols							
Amazonas	4.55	3.22	3.5	20.2	106	87	8
Quilichao	4.30	1.08	6.5	40.1	102	88	8
Caucacia	4.85	1.49	8.0	30.4	93	37	6
II. Oxisols							
Gaviotas	4.25	1.83	11.0	33.5	115	74	7
Carimagua	4.65	2.43	14.0	32.3	108	91	8
La Libertad	4.60	2.35	14.0	33.9	107	70	8
III. Andepts							
Unidad 10	5.50	3.95	36.0	41.2	59	0	5
El Refugio	5.30	0.42	57.0	72.1	88	16	7

† RAE with total superphosphate (TSP) = 100.   ‡ PR = phosphate rock.

It is justifiable to argue that the comparison shown in Table 7-2 is questionable because the properties of the soils that could affect the performance of phosphate rocks, other than P-retention capacity, were not considered, nor were these properties held constant. To remedy this situation, Hammond et al. (1986a) conducted a greenhouse experiment in which Bayovar PR and TSP were added to a soil that had previously been treated with varying amounts of amorphous Fe gel to provide a range of P-retention capacities while holding the other soil properties, such as pH, uniform. The results, shown in Fig. 7-3, demonstrated that, although both the TSP and the PR declined in effectiveness as the P-retention capacity of the soil increased, the PR tended to decline at a more rapid rate. As noted by Hammond et al. (1986b), this phenomenon has not been properly explained. One can speculate that the reduced effectiveness of the PR with increase in P-retention capacity is the result of reduced root development due to poor P supply in the early growth stages. Even though a higher P-retention capacity could result in an increase in the rate of dissolution of the PR, the concentration of P in solution would be limited to a low level controlled by the solubility product of the solid-phase apatite in the PR (Khasawneh & Doll, 1978). The P in solution thus released is also susceptible to retention by the soil in much the same way as is the P derived from TSP. The end result is that there is less and less P available to the crop as the P-retention capacity increases.

It is curious to note that myths sometimes interact to lead to correct observations. The research on direct application of phosphate rock in tropical America was initiated, for example, because it was thought that the rock would be more effective on high P-retaining soils and that the region contained a high proportion of those types of soil. The results of the research confirmed that phosphate rock is an effective P source for the region, but not for the suspected reasons. On the contrary, it is now known that the rela-

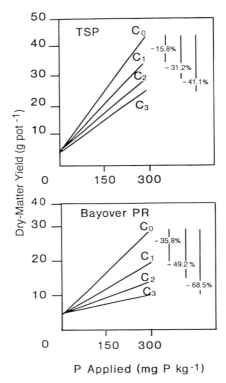

Fig. 7-3. Influence of P-retention capacity on maize response to P from triple superphosphate (TSP) and Bayovar phosphate rock (PR). Source: Hammond et al. (1986a).

tive performance of phosphate rock is maximized on soils with lower P-retention capacity, and these are the soils commonly found in tropical lowlands.

Myth or Science? Long-Term Fertilizer Use Leads to Soil Degradation

For the purpose of this chapter, soil degradation is defined as changes in the soil's chemical and physical properties that result in decreased crop production. The statement that long-term fertilizer use leads to soil degradation suggests that continuous crop production in the absence of fertilizers does not lead to soil degradation. There is ample evidence in the literature to demonstrate that this is not the case (Sanchez, 1976; Okigbo, 1984; Balasubramanian et al., 1984).

Throughout most of the tropical world, shifting cultivation has been a widely used land management system. In terms of total area, shifting cultivation is the predominant agricultural system in tropical Africa and America (Sanchez, 1976). Practiced in different forms in different regions of the tropics, shifting cultivation involves the cultivation of a piece of land for a few years followed by a fallow period when the land is allowed to return

Table 7-3. Percentage decreases in soil-fertility parameters in farmers' fields under continuous cultivation in the savanna zones of Nigeria. (Adopted from Balasubramanian et al., 1984.)

Zone	Exchangeable cations			pH
	Ca	Mg	K	
Sudan	21.0	32.0	25.0	4.0
Northern Guinea	18.6	26.8	33.0	3.8
Southern Guinea	46.0	50.6	50.0	10.0

to forest or savanna vegetation. Demographic pressure has often controlled the length of this period. Sanchez (1976) presented several reasons why farmers moved from one piece of land to the other; soil fertility depletion was singled out as the most important. Clearing and burning associated with cultivation in traditional shifting cultivation and related fallow systems cause: (i) loss of most of the N, S, and C in gases during burning and the exposure of plant nutrients to situations where they are prone to loss by erosion and leaching and (ii) destruction of humus and the deterioration of soil physical properties. The data in Fig. 7-4 and Table 7-3 demonstrate that, even under farmers' conditions, without fertilization, continuous cultivation degrades the soil.

The function of fertilizers is to promote higher crop yield per hectare. Fertilizers allow many nutrient-poor soils to become productive while at the same time enabling farmers to withdraw lands of low quality (marginal lands) from cultivation. If fertilizers are used efficiently and good farm practices are adopted, there is no evidence that fertilizers degrade the soil. Admittedly, there are numerous cases in the literature (for a complete review of the West African situation, see Pieri, 1989) where the continuous use of fertilizers has had undesirable effects on the soil (Table 7-4 and Fig. 7-5). These data show what could happen if fertilizers are used without regard to good farm management practices. For example, the data in Table 7-4 show that when mineral fertilizers are used in conjunction with manure, soil properties are maintained at desirable levels. This fact is amplified by the data in Fig. 7-6, which show the significant additive effect of mineral fertilizers and crop residue in a psammentic Alfisol in Niger. Similarly, the use of lime to correct the soil acidity that could result from the use of mineral N fertilizers can promote high yields under continuous cultivation in the sandy, poorly buffered soils of West Africa (Fig. 7-5).

The efficient use of fertilizers involves the adoption of farm practices that will allow crops to make the best possible use of added fertilizers. These practices include correct choice of the crop varieties, use of adequate plant populations, proper planting times, adequate weed and disease control, and the adoption of suitable soil conservation practices.

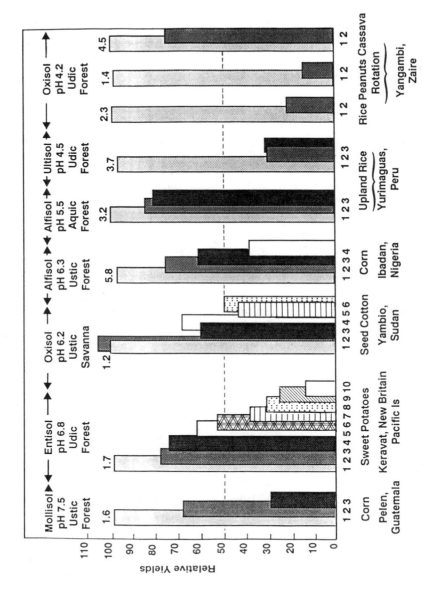

Fig. 7-4. Examples of yield declines under continuous cropping without fertilization in shifting cultivation areas as a function of soil, climate, and vegetation. (From Sanchez, 1976.)

Table 7-4. Effects of fertilizer on soil chemical properties in Burkina Faso, West Africa. Source: Abstracted from C. Pieri, 1989.

Treatment†	pH (H_2O)		Carbon		N		Ca	Mg	K	Al
	1969	1978	1969	1978	1969	1978	1978			
			g kg^{-1}				cmol$_c$ kg^{-1}			
No fertilizer	5.3	5.2	0.29	0.25	0.23	0.18	1.15	0.35	0.16	0.1
Mineral fertilizer:										
Weak dose	5.1	4.6	0.29	0.24	0.20	0.19	0.66	0.22	0.09	0.5
Strong dose	4.7	4.4	0.31	0.24	0.27	0.28	0.60	0.21	0.15	0.5
Weak dose plus manure	5.5	5.2	0.31	0.35	0.28	0.18	1.14	0.39	0.22	0.1

† Fertilizer rates: Weak dose = 18.5 kg of N/ha (1963-1970); 41.3 kg of N/ha (1971-1978); 24 kg of P_2O_5 ha^{-1}, 12.5 kg of K_2O (1963-1970); and 46 kg of P_2O_5 ha^{-1}, 18.8 kg of K_2O ha^{-1} (1971-1978). Strong dose = 50 kg of N ha^{-1}, 50 kg P_2O_5 ha^{-1}, and 12.5 kg of K_2O ha^{-1} (1963-1970); 88 kg of N ha^{-1}, 65 kg of P_2O_5 ha^{-1}, and 87.5 kg of K_2O ha$^-$ (1971-1978). Manure rate: 5 t ha^{-1}.

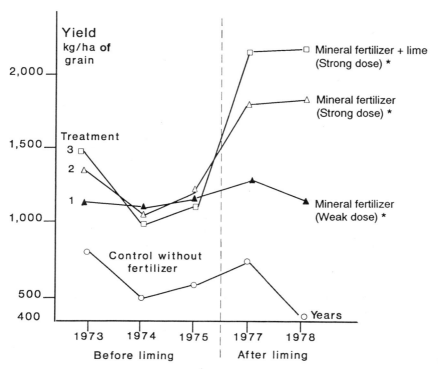

Fig. 7-5. Cotton yield obtained before and after application of soil amendment as a function of fertilizer dose, Korhogo, Cote d'Ivoire. Source: Pieri (1989).

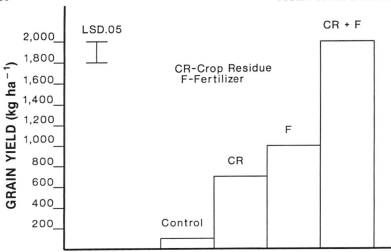

Fig. 7-6. Effect of millet grain yield response to fertilizer and crop residue application, Sadore, 1986 (IFDC, 1986).

Myth or Science? Fertilizer Use Is Too Risky Where Moisture Is Limited in the Tropics

In one form or another, water availability has a tremendous impact on agriculture in the tropics. Where lack of water is a problem, the use of a purchased input such as fertilizer is highly risk-prone. Since 1975, the West African subregion has suffered three droughts with devastating effects on agriculture and the economy of the subregion. Today, policy makers rate water availability as the major obstacle to increased agricultural production (Penning de Vries et al., 1982). Forgotten is the fact that for the past 50 years, agricultural production per hectare of cultivated land has steadily declined, even in years of excellent rainfall. In a systematic study of the actual and potential production and ecology of Sahelian rangelands, Penning de Vries et al. (1982) concluded that low soil fertility, particularly in N and P, is often the major limiting factor.

In the long run, the argument as to the relative importance of water and plant nutrients should not be left to the policy makers. The scientific fact is that the value of water to plant growth is undeniable. However, lack of water or plant nutrients results in decreased crop yields. Many practices have been developed for capturing water, but such water cannot be effectively used in an infertile soil (Ohm et al., 1985). More land is required to feed the growing population. In most cases, marginal lands are the ones brought under cultivation. Such a practice accelerates soil degradation. The International Crops Research Institute for the Semi-Arid Tropics (ICRISAT) in West Africa has demonstrated that where soil is fertile the water-use efficiency is high, resulting in increased crop production while conserving the natural resource base. Certain fertilizer nutrients such as P hasten crop maturity. In a year of early cessation of rainfall, such as happened in 1984 in West Africa, fertilizer use can mean the difference between total or partial crop failure (IFDC, 1985). Fertilizers improve land productivity. What is needed is to develop management practices, including water conservation practices, that would make fertilizer use beneficial and profitable to farmers.

Myth or Science? Because of the Risk Involved in Use of Fertilizers Under Rainfed Conditions, Small Farmers in Dry Areas Will Not Adopt Fertilizer Use

A direct result of the previous myth is the belief that farmers in semidry and dry environments who depend on only rainfall will not adopt the use of fertilizers. The International Fertilizer Development Center (IFDC) in collaboration with ICRISAT tested this belief in Gobery, a village located 100 km southeast of Niamey, in Niger.

Building on socioeconomic baseline data obtained between 1981 and 1983 by ICRISAT, covering such items as demographic characteristics of households; land tenure and use; cropping systems on household plots; farm labor and allocation of time; and fertilizer sources, use, constraints, and knowledge, IFDC initially set up a demonstration plot to illustrate the effects of fertilizers on millet (*Panicum* spp.) and cowpea [*Vigna unguiculata* (L.) Walp.]. The demonstration trial was followed in 1984 by a researcher-managed trial to measure the effect of P and N fertilizer sources and rates on crop yield. Phosphate rock from Niger was compared with single superphosphate and triple superphosphate at 0, 15, and 30 kg of P_2O_5 ha^{-1}. The P fertilizers were either broadcast or spot-placed. Results showed striking responses to P fertilizers that confirmed results obtained at the ICRISAT research station at Sadore, Niger. In 1986, 20 farmers were randomly selected to validate the information that had been obtained by the researcher in both the research station trials and the on-farm trial conducted in the village. Very few of the farmers in Gobery had used fertilizers before and their knowledge of fertilizer use was limited. During a series of meetings, farmers' opinions were elicited to develop a suitable package of fertilizers, seeds, and management practices. For these trials, farmers were provided with free fertilizers and seeds and given assistance in laying out the experimental plots. All other operations from planting to harvesting were carried out by the farmers themselves. These farmer-managed trials were repeated in 1987.

Over these 2 yr, millet yields in the farmers' fields where fertilizers were applied increased by an average of 250% (IFDC, 1986, 1987, 1988). Fertilizers improved crop establishment, crop density, and subsequent grain yield. The improved dry matter meant more crop residue for domestic use while up to 2 t ha^{-1} remained in the fields. This protected the soil from the effects of both the harsh dry season and the extremely severe storms that accompany the onset of the rains.

Although IFDC worked with only 20 of the 150 farm families in the village, a survey undertaken by ICRISAT showed that more than 98% of

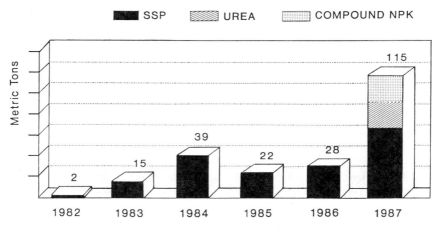

Fig. 7-7. Fertilizer use and adoption in Gobery growth in fertilizer use 1982 to 1987 (IFDC, 1988).

the farms in Gobery were fertilized in 1987. As the data in Fig. 7-7 indicate, fertilizer consumption in the village increased from about 2 t of single superphosphate (SSP) in 1982 to 115 t of SSP, urea, and compound N-P-K in 1987. Results of the baseline survey had indicated that availability and high cost of fertilizers were the major constraints to its use. The farmers organized themselves into a cooperative so they could buy fertilizers in bulk for the village and overcome the problem of fertilizer supply. This resulted in the phenomenal increase in the use of fertilizers, as depicted in Fig. 7-7, without any changes in the government's procurement or pricing policies. The yield increases meant increased food security. Food security and profitability must be considered when hazarding a guess as to what farmers can and cannot do with respect to adoption of fertilizers. The benefits of fertilizer use as seen by the farmers of Gobery must have outweighed the risk involved in fertilizer use.

SUMMARY

Tropical ecosystems are extremely complex. Lack of adequate information about the nature of these complex ecosystems has bred many myths relating to the agricultural potential of the ecosystems and about the function of fertilizers in crop production.

The highest rates of population growth are occurring in the developing countries located in the tropics. This demographic pressure is changing traditional agricultural patterns. The result is increased tendency for soil and environmental degradation. Accelerated acquisition of knowledge about tropical ecosystems and the role of fertilizers in promoting agricultural productivity while maintaining or enhancing the qualities of the environment is needed to eradicate the myths relating to fertilizer use. It is evident that in tropical ecosystems, a delicate balance exists within the soil/plant continuum. Management practices that must include efficient use of fertilizers must continue to be developed to sustain the productivity of this continuum.

REFERENCES

Baanante, C.A., B. Bumb, and T.P. Thompson. 1989. The benefits of fertilizer use in developing countries. Paper Ser. P-8. Int. Fertilizer Develop. Ctr., Muscle Shoals, AL.

Balasubramanian, V., L. Singh, L.A. Nnadi, and A.U. Mokwunye. 1984. Fertility status of some upland savanna soils of Nigeria under fallow and cultivation. Samaru J. Agric. Res. 2:13-23.

Bumb, B. 1989. Global fertilizer perspective 1960-1995. The dynamics of growth and structural change. Tech. Bull. T-34. Int. Fertilizer Develop. Ctr., Muscle Shoals, AL.

Buol, S.W., P.A. Sanchez, R.B. Catte, and M.A. Granger. 1975. Soil fertility capability classification: A technical soil classification for fertility management. p. 126-141. *In* E. Bornemisza and A. Alvarado (ed.) Soil management in tropical America. North Carolina State Univ., Raleigh.

Cochrane, T.T., and P.A. Sanchez. 1981. Land resources, soil properties and their management in the Amazon region: A state of knowledge report. *In* Conference on Amazon Land Use Research. CIAT, Cali, Colombia.

Cochrane, T.T., and P.A. Sanchez. 1982. Land resources, soils and their management in the Amazon region: A state of knowledge report. p. 137-209. *In* S. Hecht (ed.) Amazonia: Agriculture and land use research. Centro Int. Agric. Trop., Cali, Colombia.

Cooke, G.W. 1984. Phosphorus and potassium problems in plant production and how to solve them. CIEC World Fertilizer Congress, Budapest, Hungary.

Dudal, R. 1979. Soil related constraints to agricultural development in the tropics. p. 23-37. *In* Priorities for alleviating soil related constraints to food production in the tropics. Int. Rice Res. Inst., Los Baños, Philippines.

Greenland, D.J. 1973. Soil factors determining responses to phosphorus and nitrogen fertilizer used in tropical Africa. Afr. Soils 17:99-108.

Hammond, L.L., S.H. Chien, and G.W. Easterwood. 1986a. Agronomic effectiveness of Bayovar phosphate rock in soil with induced phosphorus retention. Soil Sci. Soc. Am. J. 50:1601-1606.

Hammond, L.L., S.H. Chien, and A.U. Mokwunye. 1986b. Agronomic value of unacidulated and partially acidulated phosphate rocks indigenous to the tropics. Adv. Agron. 40:89-140.

Hammond, L.L., and L.A. Leon. 1983. Relative agronomic effectiveness of some Latin American phosphate rocks in Colombian soils. p. 503-518. *In* Proc. the 3rd Int. Congr. of Phosphorus Compounds. Inst. Mondial du Phosphate, Brussels, Belgium.

International Crops Research Institute for Semi-Arid Tropics. 1985. Annual report. ICRISAT, Hyderabad, India.

International Fertilizer Development Center. 1985, 1986, 1987, 1988. Annual reports. Int. Fert. Develop. Ctr., Muscle Shoals, AL.

Khasawneh, F.E., and E.C. Doll. 1978. The use of phosphate rock for direct application to soils. Adv. Agron. 30:159-205.

Mokwunye, A., S.H. Chien, and E. Rhodes. 1986. Phosphate reactions with tropical African soils. p. 253-281. *In* A. U. Mokwunye and P.L.G. Vlek (ed.) Management of nitrogen and phosphorus fertilizers in sub-Saharan Africa. Martinus Nijhoff Publ., Dordrecht, Netherlands.

Nunez, R. 1984. Applicacion directa de la roca fosforica a suelos agricolas en Mexico. p. 265-294. *In* V. Ricaldi and S. Escalera (ed.) La Roca Fosforica. Vol. 2. Grupo Latinoamericano de investigadores de Roca fosforica (GLIRF), Cochabamba, Bolivia.

Ohm, H.W., J.G. Nagy, and S. Sawadogo. 1985. Complementary effects of tied ridging and fertilization with cultivation by manual, donkey and ox traction. *In* H.W. Ohm and J.G. Nagy (ed.) Appropriate technologies for farmers in semi-arid West Africa. Int. Programs in Agric., Purdue Univ., West Lafayette, IN.

Penning de Vries, F.W., T. Djiteye, and M.A. Djiteye (ed.). 1982. La productivite des paturages saheliens. Une etude des sols, des vegetations et de l'exploitation de cette resource naturelle. Center for Agric. Publ. and Documentation, Wageningen, Netherlands.

Pieri, C. 1989. Fertilite des terres de savannes. CIRAD-Inst. Recherches Agronomiques Tropicale, Paris.

Russel, E.W. 1974. The role of fertilizer in African agriculture. p. 213–238. *In* V. Hernandez Fermnandez (ed.) Fertilizer, crop quality and economy. Elsevier Sci. Publ. Co., New York.

Sanchez, P.A. 1976. Properties and management of soils in the tropics. John Wiley and Sons, New York.

Sanchez, P.A., and G. Uehara. 1980. Management considerations for acid soils with high phosphorus fixation capacity. p. 417–514. *In* The role of phosphorus in agriculture. ASA, Madison, WI.

Smyth, J.J., and P.A. Sanchez. 1982. Phosphate rock dissolution and availability in Cerrado soil as affected by phosphorus sorption capacity. Soil Sci. Soc. Am. J. 46:339–345.

8 Legume Response to Rhizobial Inoculation in the Tropics: Myths and Realities

P. W. Singleton, B. B. Bohlool, and P. L. Nakao

University of Hawaii
Paia, Hawaii

On the global scale, symbiotic N_2-fixing legumes contribute a major source of N to the biosphere, though estimates of the actual amounts vary considerably (e.g., see Evans & Barber, 1977). In the tropics, legumes are an integral part of forest ecosystems, natural pastures, and agricultural systems. Legume biological N_2 fixation (BNF) offers an especially attractive alternative to fossil fuel-derived N fertilizers in subsistence farming and sustainable agricultural programs in developing countries where hard currency is in short supply and fertilizer N is virtually unavailable on the local markets.

In the USA, Canada, Australia, and other developed countries, demonstration of benefits from inoculating certain legumes has led to the formation of commercial enterprises based on the inoculant production technology. In the developing tropics, however, promotion and use of rhizobial inoculants and the establishment of an inoculant industry would be successful only if significant benefits could be demonstrated for important tropical legumes.

Inoculation success/failure is highly site specific. It depends on such a multitude of interacting factors that it would be virtually impossible to make generalizations based on data gathered by different workers without standardization. Despite these difficulties, Vincent (1982) has done an admirable job of compiling information gathered from scientists from South and Southeast Asia participating in a workshop, Nitrogen Fixation by Legumes for Tropical Agriculture. Much of the information is sketchy and incomplete, but as the author points out ". . .together they give a useful insight into the significance of legumes, needs associated with their successful utilization and some indications of potential for an improved contribution."

Nutman (1976) describes a series of inoculation trials conducted under the organization of the International Biology Program (IBP) during the late 1960s and early 1970s. The design of these experiments included measures of factors controlling the response to inoculation of alfalfa (*Medicago sativa* L.) in numerous locations. While the benefit of inoculating alfalfa was clearly demonstrated in these experiments, there has not been until recent

---

Copyright © 1992 Soil Science Society of America and American Society of Agronomy, 677 S. Segoe Rd., Madison, WI 53711, USA. *Myths and Science of Soils of the Tropics.* SSSA Special Publication no. 29.

years a comprehensive effort to evaluate the response to rhizobial inoculation of tropical legumes. Given the widespread misconceptions about tropical rhizobia and the need for farmers in the tropics to inoculate their legume crops, it is imperative that reliable data on the benefits from inoculation be available.

THE MYTH OF THE PROMISCUITY OF TROPICAL LEGUMES

Legumes require the aid of their symbiotic nodule-forming bacterial partners, rhizobia, for N_2 fixation. Not all rhizobia are capable of forming nodules on all legumes. There are host-bacterial specificities expressed by both partners. The so-called "cross-inoculation" groups are comprised of groups of legumes that nodulate with the same group of bacteria, which are then classified together in the same bacterial species. Some cross-inoculation groups are highly specific and contain a limited number of legume species, while others are more promiscuous and consist of a wide range of legumes. From the bacterial perspective, some species are highly selective and others have a broader host range. This host range serves the basis for the classification of the root nodule bacteria given in Table 8-1.

The genus *Rhizobium* contains all of the "fast-growing" acid-producing bacteria, while the genus *Bradyrhizobium* contains those that are slow growing and do not produce acid on yeast-mannitol media. The categories *R. loti* and *Bradyrhizobium* spp. are basically "dumping-grounds" for groups of related and not-so-related bacteria whose host ranges are not fully known. The *Bradyrhizobium* spp. or "cowpea miscellany," for example, are purported to contain the rhizobia that nodulate tropical legumes, and here lies

Table 8-1. Summary of recent modifications to rhizobial classification: Family Rhizobiaceae.[†]

Bacteria	Host legume
Genus: *Rhizobium*	
R. leguminosarum	
bv. *viceae*	*Vicia/Pisum/Lens/Lathyrus*
bv. *trifolii*	*Trifolium* spp. (with exceptions)
bv. *phaseoli*	*Phaseolus* spp. (temperate species *vulgaris, angustifolius, coccineus*)
R. meliloti	*Melilotus/Medicago/Trigonella*
R. loti	*Lotus/Lupinus/Cicer/Anthyllis/Leucaena* and many other tropical tree legumes
R. fredii[‡]	*Glycine* spp.
Genus: *Azorhizobium*	
A. caulinodans	*Sesbania rostrata* (stem nodules)
Genus: *Bradyrhizobium*	
B. japonicum	*Glycine* spp. (*G. max, G. soja*)
Bradyrhizobium sp.	Many tropical legumes

[†] Compiled from Dreyfus et al. (1988), Jordan (1984), Scholla and Elkan (1984).
[‡] It has recently been proposed that the *R. fredii* group be renamed *Sinorhizobium fredii* comb. nov. and, along with *S. xinjiangensis* sp. nov., be placed in a separate genus *Sinorhizobium* gen. nov. (Chen et al., 1988).

the basis for the myth that because of their promiscuity, tropical legumes do not respond to inoculation in the field.

What is most important to recognize in dealing with this myth is that taxonomic units are defined by scientists for their own convenience. Also, bacteria do not recognize taxonomic boundaries; they are limited only by their own genetic compositions. The limitations of the taxonomic classifications are underscored by the fact that there are cross-overs in terms of host nodulation between bacteria in different species as well as genera. For example, it is now well known that soybean [*Glycine max* (L.) Merr.], lupin (*Lupinus* spp.), lotus (*Lotus* spp.), and others can be nodulated by bacteria in both the *Rhizobium* and *Bradyrhizobium* genera.

It is equally important to recognize that within each "infectivity" grouping there are both effectivity and infectivity subgroups. Not all bacterial strains nodulating a group of legumes are effective in N_2 fixation on all the species. An example is the case of *R. leguminosarum* bv. *trifolii* that nodulate clover species. Not all strains that are highly effective on red clover (*Trifolium pratense* L.) and white clover (*T. repens*) are necessarily effective on cultivars of subclover (*T. subterraneum* L.) (Gibson et al., 1975). Burton (1979) compiled a comprehensive list of "effectiveness groupings" for several agriculturally important legumes within cross-inoculation groups. But, he cautions that not all plants in an effectiveness group will respond similarly to all the rhizobial strains.

Graham and Hubbell (1975) have grouped tropical legumes into three broad categories based on their nodulation and N_2-fixation specificity. The group containing *Vigna, Calopogonium, Arachis, Dolichos, Glycine wightii*, and others are promiscuous and nodulate with an array of bacteria isolated from a wide range of legumes. The second group contains legumes that are highly specific in their rhizobial requirement. Such legumes as *Trifolium semipilosum* Fresen., *Glycine max (L.), Leucaena leucocephala* (Lam.), and *Lotononis bainesii* are placed in this group. Intermediate between the two extremes is the third group of tropical legumes that nodulate with a wide range of rhizobia, but often ineffectively. Such legumes as *Centrosema* spp., *Desmodium*, some *Stylosanthes* spp., and *Phaseolus vulgaris* L. belong to this group. More recently, Date (1982) has proposed a similar specificity classification scheme for tropical forage legumes that include three categories: (i) Group PE, Promiscuous and Effective; (ii) Group PI, Promiscuous and Ineffective; and (iii) Group S, Specific.

Even among the promiscuous group, there are those that nodulate with a subgroup of bacteria. Earlier, Doku (1969) observed that all bacteria derived from nodules of peanut (*Arachis hypogaea* L.) and Bambarra groundnut [*Vigna subterranea* (L.)] were capable of nodulating each other as well as cowpea [*V. unguiculata* (L.)] and lima bean (*Phaseolus lunatus* L.). However, not all isolates nodulating cowpea were able to nodulate peanut and Bambarra groundnut. In addition, Ahmad et al. (1981) showed that not all of the West African isolates from cowpea, a promiscuous legume which is thought not to require rhizobial inoculation in tropical soils (Sellschop,

1962), were capable of nodulating peanut, pigeonpea, or mungbean [*V. radiata* (L.)], which are all classified in the "cowpea miscellany" cross-inoculation group.

THE MYTH OF THE LACK OF LEGUME RESPONSE TO INOCULATION IN THE TROPICS

International Network of Legume Inoculation Trials

The myth that tropical legumes do not need inoculation, and the lack of hard evidence to the contrary, had created a void in the supply of, and demand for, inoculants in most developing countries. In 1979, Univ. of Hawaii's Nitrogen Fixation by Tropical Agricultural Legumes (NifTAL) Project organized a planning workshop to develop an experimental protocol and strategy for implementation of standardized inoculation trials on a global scale (Harris, 1979). Twenty-nine agricultural, biological and social scientists, biometricians, administrators, and management specialists worked together to develop appropriate experimental designs and procedures to be implemented by volunteer cooperators in developing countries. The primary purpose of this 5-yr effort was to determine under realistic field conditions whether yield of agriculturally important tropical legumes could be enhanced by rhizobial inoculation. An added benefit was that of demonstrating the potential usefulness of BNF technology for agricultural development to farmers, extension workers, scientists, and decision makers. The scope of International Network of Legume Inoculation Trials (INLIT) encompassed a standard set of experiments to test inoculation response of legumes under two levels of fertility (farmers' local practices and maximal level), and in comparison with added fertilizer N. INLIT cooperators conducted more than 200 standardized experiments with 19 species of legume field sites in more than 20 countries.

The treatments, originally designed by the INLIT Workshop participants (Harris, 1979), were considered the minimum that needed to be done to address the objectives in a realistic and statistically reliable manner. Treatments for the INLIT trials included three N-source treatments: (i) inoculated; (ii) uninoculated; and (iii) fertilizer N. These N-source treatments were conducted at two levels of management: (i) farmer conditions; and (ii) maximal management (for details of design and analysis see Davis et al., 1985). All cooperators were urged to adhere to standardized guidelines for plot layout, dimensions, and planting information.

A cumulative summary of the data from INLIT is given in Table 8-2. In a large number of trials, a significant increase (>1.0 SD) was obtained under both fertility levels. Even legumes that belong to the "cowpea miscellany" cross-inoculation group (*A. hypogea, C. cajan, V. mungo, V. radiata,* and *V. unguiculata*) had significantly higher yields with rhizobial inoculation in a large percentage of the trials. In 29% of the cases, improving other fertility factors increased the chances of inoculation success. Without

Table 8-2. Yield response of tropical legumes to rhizobial inoculation: Summary of results.†

		A	B	Significant response to inoculation, % of total‡		
				Fertility level		Man-age-ment
Scientific name	Common name	Total trials	% A, CV <20%	Farm	Max.	
Arachis hypogaea L.	Peanut	26	88	50	46	23
Cicer arietinum L.	Chickpea	31	61	48	55	23
Cajanus cajan (L.)	Pigeonpea	8	38	13	13	88
Glycine max (L.)	Soybean	40	62	65	65	30
Lens culinaris Medikus	Lentil	27	78	48	41	11
Leucaena leucocephala (Lam.)	Leucaena	8	63	38	50	25
Phaseolus vulgaris L.	Common bean	10	80	10	30	50
Vigna mungo (L.)	Gram, black	15	67	53	60	47
Vigna radiata (L.)	Mung bean	40	60	70	68	28
Vigna unguiculata (L.)	Cowpea	9	56	56	11	11
Other§		14	71	36	50	21
Total		228				

† Compiled from data of cooperators of the International Network of Legume Inoculation Trials (INLIT), sponsored by NifTAL Project, coordinated by R.J. Davis. Column A = No. of trials acceptable out of those analyzed; B = percent of A with CV <20%.
‡ Percent of column A, with yield of inoculated greater than uninoculated by greater than SD. Management refers to the percent of trials in which inoculated treatments at max. were significantly greater than those at farm fertility.
§ "Other" includes: *Aeschynomene americana* (1), *Calopogonium caeru* (1), *Centrosema pubescense* (1), *Medicago sativa* (1), *Pisum sativum* (5), *Psophocarpus tetragonolobus* (2), *Pueraris phaseoloides* (1), *Sesbania sesban* (1), and *Vicia faba* (1).

additional environmental information, it is difficult to explain why some cooperators obtained a positive inoculation response at farm fertility levels, but not at maximal fertility. Such are the vagaries of field trials conducted on different species by different individuals at different sites.

A breakdown of the INLIT results by country (Table 8-3) shows that legumes that respond to inoculation in one country do not necessarily do so in another. The pattern of response, however, does not correspond strictly to the center of origin and diversity of the legume species. This variability indicates that there is enough local heterogeneity and site specificity to account for a legume species responding to inoculation in its own center of origin. It should be noted that some of the trials were conducted at government and university experimental stations and, therefore, were affected by prior experimental agricultural practices that may or may not have included rhizobial inoculation.

Measuring the Response of Legume Inoculation: Limitations of Standard Inoculation Trials

Recommending that farmers inoculate or that national programs and private industry invest resources to enhance the delivery of BNF technology is predicated on the results of inoculation trials. INLIT demonstrated that

Table 8-3. Distribution of yield response of specific legumes by country at two levels of fertility.

Legume species	Total no.	Response at fertility† Farm	Response at fertility† Max.	Legume species	Total no.	Response at fertility† Farm	Response at fertility† Max.
Arachis hypogaea L				*Leucaena leucocephala* (Lam.)			
Argentina	1	1	0	Brazil	1	1	1
Burma	1	0	0	India	2	0	0
Ghana	2	2	0	Mexico	1	0	1
India	6	2	2	Philippines	2	2	1
Indonesia	4	2	2	Puerto Rico	1	0	1
Malaysia	1	0	0	USA	1	0	0
Philippines	9	5	6	Total	8	3	4
Rwanda	1	0	1	*Phaseolus vulgaris* L.			
Vietnam	1	1	1	Brazil	1	0	0
Total	26	13	12	Cameroon	1	0	0
Cajanus cajan (L.)				Ethiopia	3	0	2
Fiji	3	1	0	Guatemala	1	0	0
India	5	0	1	India	1	0	0
Total	8	1	1	Indonesia	1	1	1
Cicer arietinum L.				Mexico	1	0	0
Bangladesh	3	3	3	USA	1	0	0
Burma	1	0	0	Total	10	1	3
India	22	10	11	*Pisum sativum* L.			
Nepal	1	0	0	Guatemala	1	0	1
Pakistan	3	1	1	India	4	2	2
USA	1	1	0	Total	5	2	3
Total	31	15	15	*Vigna Mango* (L.)			
Glycine mas (L.)				Bangladesh	1	0	0
Colombia	1	0	0	India	14	8	9
Egypt	1	1	0	Total	15	8	9
Ethiopia	2	0	1	*V. radiata* L.			
India	2	1	1	Bangladesh	1	0	0
Indonesia	4	2	1	Burma	1	0	0
Malaysia	2	1	2	Fiji	1	0	0
Mauritius	1	1	1	India	23	17	17
Nepal	2	0	0	Pakistan	3	2	2
Pakistan	3	3	3	Philippines	10	9	7
Philippines	9	7	6	Swaziland	1	0	1
Rwanda	1	1	1	Total	40	28	27
Sri Lanka	1	0	0	*V. unguiculata* (L.)			
Sudan	1	1	1	Fiji	2	0	0
Taiwan	1	1	1	India	6	2	1
Thailand	1	1	1	Rwanda	1	0	0
USA	1	1	1	Total	9	2	1
Vietnam	3	2	3				
Total	36	23	23				
Lens culinaris Medikus							
Bangladesh	1	0	1				
India	22	11	10				
Nepal	2	2	0				
Pakistan	1	0	0				
USA	1	0	0				
Total	27	13	11				

† Number of experiments with yield of inoculated greater than uninoculated by at least 1.0 SD. Results from USA are from INLIT trials conducted in Hawaii.

Table 8-4. Frequency of inoculation response and increase in seed protein content. (From J. Thies, Ph.D. thesis, Univ. of Hawaii, 1990.)

Legume protein	Inoculation increased		Average seed N	
	Yield	Seed N	+ inoc.	− Inoc.
	% of trials		g N kg^{-1}	
Soybean	83	100	38.8	35.6
Lima bean	60	80	19.4	18.8
Common bean	33	50	18.8	17.5
Cowpea	0	80	26.3	24.4

the yield response to legume inoculation is highly variable and can be difficult to measure statistically with small plot field experiments. INLIT results suggest that other criteria for measuring the response to legume inoculation and standards for accepting results should be considered.

Limitations of Measuring Yield Response

The benefit to inoculation must be considered in the context of the small investment required to inoculate legumes. Table 8-4 shows that inoculation increased seed protein content more frequently than yield. While yield is a major criterion, better seed quality, total N increases, or conservation of soil N should also be considered as criteria for a positive result from inoculation trials.

It makes little sense that experimental results meet rigid statistical tests prior to recommending inoculation to the farmer. The variability inherent in even well-managed small plot field experiments is large. Only under the best experimental management are coefficients of variation (CV) <15% obtained. This variation precludes measuring statistically significant results of any single experiment when, depending on average yield, the yield increase is <300 kg ha^{-1}. Given the low cost of inoculant, measuring an economically relevant response to legume inoculation in terms of yield is limited when traditional statistical tests are applied to field experiments. New methods of measurement and predicting benefits from legume inoculation are clearly indicated if BNF researchers are to make valid recommendations on the need to develop and deliver inoculation technology to farmers.

The Site-Specific Nature of Response to Inoculation Trials

Although field experimentation is a standard technique for measuring the response to inoculation, the manner in which inoculation trials have been performed in the past preclude obtaining results of relevance to locations other than the site where they were performed. Most trials were conducted without a measure of any factor that may affect or correlate with the response to inoculation. The results of these individuals trials and reasons for inoculation response remain site specific in both time and space. Using the results of these trials to project the performance of inoculant at other sites assumes that environmental, soil, and biological factors do not play a role in determining the response to legume inoculation.

While many trials were designed to determine legume response to inoculation under local conditions, others had regional or global perspectives. Programs such as INLIT (Harris, 1979), or those sponsored by IPB (Nutman, 1976) and International Center of Tropical Agriculture (CIAT), Cali, Columbia, were intended to demonstrate the need for inoculant with numerous crops in a variety of environments using standard experimental protocols. These programs were effective in demonstrating that some legumes responded to inoculation in some environments. Compared to single location trials, the results of these programs offer more information and can be used to generate a probability statement of expected response to inoculation by a legume over the range of conditions included in the trials. The usefulness of this information is, however, restricted since the soil and climate conditions of the trials are not specified sufficiently to identify particular situations where a response is likely. Neither planning agencies, industry, nor farmers and extension workers can confidently use this information as criteria to develop or promote inoculation technology under their specific conditions.

Understanding the factors that regulate the response to legume inoculation may lead to quantitative models that can predict the need for, and the benefit from, legume inoculation in any environment. Following is a discussion on factors regulating the response to inoculation and some experimental results that indicate a new approach for determining whether farmers are likely to benefit from inoculation.

A New Approach to Inoculation Trials: The Worldwide Rhizobial Ecology Network

The Worldwide Rhizobial Ecology Network (WREN) was established with the hypothesis that measurable environmental factors determine and could be used to predict the performance of rhizobial inoculant in tropical soils. Similar to INLIT and other inoculation programs, the WREN developed a standardized set of experimental protocols to measure the response to inoculation. A network of collaborators was established in 17 countries in the tropics. Collaborators were selected on the basis of their experience with earlier networks and rhizobial technology.

WREN is unique in that the site specificity of inoculation trials was formally addressed. Experimental protocols included measures of factors most influencing the response to inoculation. Specific protocols included in the design of WREN experiments include: (i) a non-N_2-fixing legume species to evaluate the availability of soil N; (ii) a fertilizer N control providing N in excess of that required for growth to measure yield potential; and (iii) evaluation of several legume species with different rhizobial requirements at the same site. A database of indigenous rhizobial numbers and soil and environmental parameters was developed for each site.

While many parameters were measured, there were three which, when combined, explained more than 80% of the variation between results of the inoculation trials. These factors are: (i) yield potential; (ii) soil N availability; and (iii) the size of the indigenous population of rhizobia. The most important of the three factors is the number of rhizobia in the soil at planting. Mathematical models describing the relation between these factors and the response to inoculation have been reported (Thies et al., 1991a, b).

Regulation of the Response to Legume Inoculation: A Conceptual Model

A conceptual framework of the factors influencing the response to legume inoculation is presented in Fig. 8-1. Legumes have two sources from which to assimilate N: (i) mineral N from the soil or fertilizer and (ii) the atmosphere. Total potential N assimilation (Crop Demand for Nitrogen) occurs in proportion to growth, and is determined by a complement of soil and environmental factors in the system. Mineral N availability above small quantities in early seedling growth reduces assimilation of atmospheric N (George et al., 1988; Herridge & Brockwell, 1988; Imsande, 1986; Ralston & Imsande, 1983) and regulates the crop N deficit that must be met through BNF. Therefore, the relationship between legume-N requirement and soil-N availability defines the potential amount of atmospheric N required by the legume.

The crop N deficit can be met through BNF only if there are sufficient quality and quantity of rhizobia available from the soil and inoculant sources. The potential crop response to inoculation is then defined by the quality of the indigenous rhizobia in the soil at inoculation. When populations of rhizobia in the soil are insufficient to meet the demand for fixed N by the legume, a response to inoculation will be observed. The important variables in this conceptual framework are addressed below. It is certain that unless there is a quantitative understanding of these regulating variables and their interaction, further inoculation trials will continue to yield the same site-specific information as the multitudes of earlier trials.

Factors Affecting the Response to Inoculation

Yield Potential of Legumes and the Response to Inoculation. Environmental and management variables influence legume yield and, as a result, the requirement for atmospheric N. Nitrogen fixation and the potential response to inoculation, therefore, are also necessarily affected by the en-

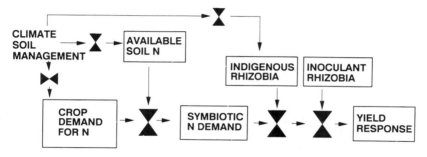

Fig. 8-1. Conceptual model of the regulation of the response to legume inoculation.

vironment. Management and environmental considerations have received little attention in relation to N_2 fixation and the response to legume inoculation with rhizobia. Based on the conceptual framework presented in Fig. 8-1, the influence of management and environment on the potential response to legume inoculation is a function of their relative impacts on plant growth (crop N demand), rhizobia, and symbiotic processes (N supply).

Despite the many symbiotic processes that may influence the response to inoculation, most research to improve symbiotic performance in the field has focused on strain selection for effectiveness (Boonkerd et al., 1978; Ham et al., 1971; Harris, 1979) and environmental influences on rhizobia as free-living organisms (for a comprehensive review see Lowendorf, 1980). Significant effort has been directed toward selecting rhizobia for tolerance to environmental stress (Cassman et al., 1981a, b; Keyser & Munns, 1979) with the goal of improving symbiotic performance in environments with ecological constraints to legume productivity. While the saprophytic phase is an important determinant of the population size of indigenous rhizobia and survival of rhizobia in inoculant, it is often not limiting the response to legume inoculation. Rhizobia are generally more tolerant than the host plant to many soil stresses.

Figure 8-2 shows the result of an inoculation trial with soybean using two strains of *Bradyrhizobium japonicum* that had been shown to differ in ability to grow in cultures low in available P (Cassman et al., 1981a, b). Strain USDA 110 performed better in vitro under P-limited conditions than strain

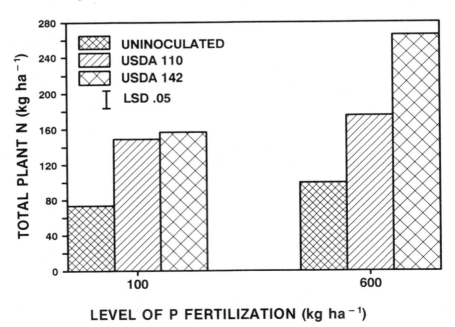

Fig. 8-2. Effect of P and strain of *Bradyrhizobium japonicum* on the response to inoculation of soybean.

USDA 142. Comparing the performance of the two strains in a P-deficient soil indicates that the P status of the soil regulated the response to inoculation (R. Nyemba, M.Sc. thesis, Univ. of Hawaii, 1986). Strain USDA 142 performed as well as USDA 110 under P-limited conditions but fixed significantly more N when additional P was added. Differences between strains were not large until the P constraint was completely removed, indicating that BNF effectiveness was fully expressed only when crop N demand was enhanced by P fertilization. Similar conclusions regarding P availability and rhizobial performance have been reached in controlled greenhouse experiments (Singleton et al., 1985).

There have been few reports that identify which stages of the symbiotic process are most sensitive to environmental stresses. The lack of research in this area is due to the difficulty in isolating the effect of environment on particular processes. Various approaches have been used to identify the first limiting process of the symbiosis in conditions of environmental stress. The approaches have either isolated the effects of environment on nodule formation or function in time (Munns, 1968), space (Hinson, 1975; Singleton & Bohlool, 1983, 1984), or by enhancing the supply of photosynthate supply in conditions of soil moisture deficit through enhanced CO_2 concentration (Huang et al., 1975).

Results from these limited number of studies indicate that soil stress factors such as drought (Huang, 1975) and salinity (Singleton & Bohlool, 1983) did not directly impact nodule function. The effects of these soil stress factors on an existing symbiosis was through a reduction in growth and photosynthesis. Nodule formation has been shown to be more sensitive to soil acidity (Munns, 1968) and salinity (Singleton & Bohlool, 1983) than plant growth of alfalfa and soybean. The presence of mineral N does not reduce plant growth and affects the infection process more than nodule function (Hinson, 1975). While limited, these reports indicate that under some agricultural environments, the magnitude of response to inoculation will be mainly influenced by general crop adaptation rather than specific direct impacts of environment on symbiotic processes.

Soil Nitrogen Availability. The difference between soil N available to the crop and crop N requirements determines the N deficit that must be met by BNF. While many agricultural soils cannot supply N in sufficient quantities to eliminate the need for BNF, the amount of N available can be a major factor contributing to the magnitude of the response to inoculation.

Many reports exist on the effects of mineral N on nodulation and N_2 fixation by legumes (see Munns, 1977, for review). It is generally accepted that small quantities of mineral N available during early growth can promote nodulation and N_2 fixation. This increase is most likely due to increased plant vigor during the establishment of symbiotic structures. Additional available mineral N substitutes for N_2 fixation, and if the mineral N is supplied in abundance, it can completely suppress the symbiosis (Gibson & Harper, 1985; Ismande, 1986).

Table 8-5. The effect of potential crop N deficit on soybean response to inoculation. (Data from George, 1988.)

Crop N deficit[†]	Inoculation response
kg ha^{-1}	%
0	5.8
7	2.4
16	11.0
19	16.1
131	37.9
206	113.0

[†] Crop N deficit calculated by subtracting N assimilated by a nonnod soybean isoline from N assimilated by inoculated Clark soybean supplied with a total of 900 kg of N applied in three equal applications at planting, R2/R3, and R4/R5.

There are few reports that quantify the effect of mineral N on N_2 fixation and productivity in the field. Weber (1966) demonstrated that mineral N regulated the symbiosis of field-grown soybean over a 7-yr period. Nitrogen applied at rates up to 600 kg of N ha^{-1} was required to completely suppress N_2 fixation. In years with reduced rainfall, N_2 fixation was eliminated with as little as 150 kg of N ha^{-1}.

Data in Table 8-5 demonstrates how the magnitude of crop N deficit affects the potenital response to inoculation of soybean. Crop N deficit for data in Table 8-5 (George, 1988) was calculated by subtracting the N assimilated by a nonnodulating soybean isoline from that assimilated by inoculated soybean supplied with 900 kg of N ha^{-1}. The lack of significant nodulation by the plus N control indicated that this treatment was measuring near-maximum N assimilation potential. The nonfixing soybean measured total soil N availability. The crop N deficit in effect integrates both soil N availability and yield potential in the system.

Populations of Rhizobia. Although the quality and size of indigenous populations of rhizobia regulates the response to inoculation, there are few reports where measures of rhizobial populations are correlated with crop performance. Indigenous populations of rhizobia are probably the single most important, but least studied, of any factor that conditions the response to legume inoculation.

The indigenous rhizobia in the soil present a barrier to increasing crop yield through inoculation. Native rhizobia compete with inoculant strains for nodulation of the host (Berg et al., 1988; Bohlool & Schmidt, 1973; Weaver & Frederick, 1974a, b). Indigenous populations of rhizobia also may be highly effective at N_2 fixation (Brockwell et al., 1988; Gibson et al., 1975; Singleton & Tavares, 1986). Following is a discussion of rhizobial populations in tropical soils, and their influence on the success of inoculum strains and response to legume inoculation.

Table 8-6. Frequency distribution of rhizobial population densities in tropical soils.†

Number rhizobia‡	Test host				
	Soybean	Cowpea§	Lima bean	Common bean	*Leucaena leucocephala*
g^{-1} soil			% of samples		
0	47	16	26	35	51
1-10	13	9	26	11	9
10-100	15	21	21	19	15
100-1000	10	12	5	15	9
>1000	14	40	21	19	15
No. of samples	105	122	19	26	33

† Data from WREN: Ecuador (G. Bernal; C. Estevez); Costa Rica (C. Ramirez); Philippines (E. Paterno); Egypt (H. Moawad); Honduras (J. Rosas); India (S.V. Hegde); Morocco (A. Hilali); Pakistan (K. Malik; F. Yusef); Ghana (R. Abaidoo); Puerto Rico (E. Schroeder); Taiwan (C.C. Young); Zimbabwe (M. Nyika); Zambia (R. Nyemba; V. Chinene); Brazil (W. Ribeiro); Thailand (N. Boonkerd; P. Wadisirisuk); Hawaii (Theis, Ph.D. thesis, Univ. of Hawaii, 1990; Woomer et al., 1988); Guyana (Trotman & Weaver, 1986); Indonesia (P. Singleton, B. Hilton, S. Saroso, and N. Boonkerd, 1990, unpublished data); and India (P. Singleton and S.V. Hegde, 1990, unpublished data).
‡ Zero indicates rhizobial numbers were less than the detection limit of 0.4 rhizobia g^{-1} soil.
§ Hosts include: cowpea, mung bean, groundnut, and *Macroptilium atropurpureum* L.).

THE MYTH OF THE UBIQUITY AND ABUNDANCE OF TROPICAL RHIZOBIA

Misconceptions about distribution of tropical rhizobia stem mostly from the confusion created by the taxonomic units to which these rhizobia are assigned, that is, cowpea miscellany, as discussed earlier. Data from 305 soil samples in the tropics (Table 8-6) illustrate that while rhizobia for cowpea and a few other promiscuous legumes are present in a large number of sites, the population densities are extremely variable. Almost one-half of the soils sampled had fewer than 100 rhizobia g^{-1} soil. The distribution of rhizobia of legumes of the subgroups of the cowpea miscellany (e.g., *P. lunatus*) indicate that more soils have low numbers of rhizobia for these legumes. It simply cannot be assumed that tropical soils have sufficient numbers of *Bradyrhizobium* spp. to meet crop N demand.

Rhizobia for several of the important legumes grown in the tropics (i.e., *L. leucocephala* and *G. max*) are either not present at all in many of the locations, or occur in such low numbers that would necessitate inoculation.

Ecological Factors Influencing the Indigenous Populations of Rhizobia

The presence and size of indigenous populations of rhizobia is a function of climate, soil, crop history, and management (Lawson et al., 1987; Weaver et al., 1987; Woomer et al., 1988; Yousef et al., 1987). Woomer et al. (1988) developed mathematical expressions describing the population density of six species of rhizobia in tropical soils in terms of climate, soil, and legume cover while Weaver et al. (1987) demonstrated the importance of crop history, and Yousef et al. (1987) described the importance of crop history and soil factors. Species of rhizobia were restricted to ecosystems where a homologous host grows. Rainfall, soil chemical properties, and the intensity of the legume component in the natural vegetation or crop system have been identified as the most important factors determining the size of rhizobial populations in soil (Woomer et al., 1988).

Competition for Nodule Sites by Indigenous Rhizobia

Numerous reports indicate that competition from native strains of homologous rhizobia represent a constraint to establishing the inoculant on the root system and obtaining a response to legume inoculation. Strains differ in their competitive ability both in soil and artificial media (see Parker et al., 1977 for a review). Weaver and Frederick (1974a, b) and Bohlool and Schmidt (1973) concluded that success in establishing inoculant strains of rhizobia in nodules of soybean was a function of the population size in the soil. Others have concluded that competition is a major factor affecting the response to legume inoculation (Boonkerd et al., 1978; Johnson et al., 1965; Weaver & Frederick, 1974b).

A major criterion for selecting strains of rhizobia is that they will compete effectively with indigenous rhizobia to form nodules on the inoculated crop. The widespread distribution of *Bradyrhizobium* spp. in the tropics suggests competition is a major constraint to establishing inoculant strains in tropical soils. Little information is available on the conditions that affect competition between rhizobia in tropical soils. Table 8-7 demonstrates that inoculation increases the number of nodules on legumes growing in tropical soils. These data indicate that although competition is a problem in tropical soils, current inoculation technology can successfully establish superior strains of rhizobia on the root systems of tropical legumes even when there are significant numbers of indigenous rhizobia in the soil. It is doubtful that competition alone explains the failure of tropical legumes to respond to inoculation when quality inoculant and appropriate application methodology is practiced.

Table 8-7. Relative increase in nodule number from inoculation of legumes in soils with indigenous populations of rhizobia.†

Indigenous rhizobia‡	Nodule no. increase	No. observations
no. g^{-1} soil	%	no.
1–1000	434	12
>1000	155	15

† Data from WREN: Egypt (H. Moawad); Ghana (R. Abaidoo); Thailand (N. Boonkerd; P. Wadisirisuk); Taiwan (C.C. Young); India (S.V. Hegde); Morocco (A. Hilali); Ecuador (C. Estevez; G. Bernal).
‡ Rates of inoculation range from 10^6 to 10^7 rhizobia per seed.

The Size of Indigenous Populations of Rhizobia: Impact on Response to Inoculation

The symbiotic potential of populations of rhizobia can be defined by their number and effectiveness at N_2 fixation with the host. The conceptual model indicates that if the crop demand for symbiotic N is met by the indigenous population, then there can be no increase in yield and N_2 fixation through inoculation. Brockwell (1977) concluded that although numerous surveys of naturally occurring rhizobia had been conducted, "investigations of occurrence, frequency, and effectiveness of rhizobia as a means of establishing or justifying the need for legume seed inoculation have been far less frequent." While passing reference to the presence of effective native rhizobia has been used to explain the failure of legumes to respond to inoculation (Diatloff & Langford, 1975; Ham et al., 1971; Meade et al., 1985), there are only a few reports that have attempted to measure the effect of indigenous rhizobia on the response to inoculation (Weaver & Frederick, 1974b; Singleton & Tavares, 1986; Thies et al., 1991a).

In a series of inoculation trials with soybean, Weaver and Frederick (1974b) did not observe a significant response to inoculation of soybean in six soils with indigenous *Bradyrhizobium japonicum* populations ranging from 11 to 229 086 rhizobia g^{-1} of soil. Failure to respond to inoculation when there were even low numbers of rhizobia in the soil was attributed to soil N availability and competition from indigenous strains of rhizobia, even though the inoculant rhizobia formed more than 50% of the nodules when populations of indigenous rhizobia were below 10 000 per gram of soil.

Under more controlled experimental conditions, Singleton and Tavares (1986) isolated the effect of indigenous population size on the response to inoculation by eliminating soil N availability and competition from indigenous rhizobia as experimental variables. They did not measure significant increases in N_2 fixation by six legume species when indigenous homologous rhizobia were above 20 per gram of soil. Results of these experiments are presented in Fig. 8-3. Variation in relative response to inoculation when there are no indigenous rhizobia represents environmental impact on yield potential (in this case, the influence of time of year the experiment was conducted). Extensive field experimentation has recently confirmed this relationship between numbers of rhizobia in the soil and the response to inoculation (Thies et al., 1991a). From these experiments, quantitative models based on the number of indigenous rhizobia, soil N availability and the crop N deficit have been developed to describe the response to inoculation (Thies et al., 1991b). These models were developed with eight legume species including five grains, three pasture, and one tree legume.

It is surprising that such low population densities of rhizobia can support maximum N_2 fixation by legumes, and that N_2 fixation is so sensitive to small changes in numbers of rhizobia. The nonlinear relationship between the relative response to inoculation and population density of rhizobia in the soil at planting may be explained by three possibilities.

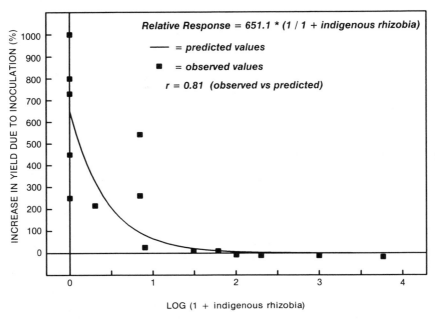

Fig. 8-3. Response to legume inoculation in relation to numbers of indigenous rhizobia: Greenhouse studies.

First, colonization of the roots and rhizosphere by rhizobia may play a role in increasing nodule number when there are few rhizobia in the soil. This mechanism has never been tested and is probably not significant. Otherwise, inoculation would not enhance nodule number as consistently as indicated in Table 8-7.

Second, even when population densities in the soil are low the root quickly exploits large volumes of soil, thereby increasing the total number of potential contacts with indigenous rhizobia. With populations of only 100 rhizobia g^{-1} of soil, there are more rhizobia in 10 kg of soil than would be applied to a large-seeded legume through inoculation.

Third, the host compensates for low numbers of nodules formed on the plant when rhizobia in the soil are few in number by increasing the size of nodules (Singleton & Stockinger, 1983). The average size of soybean nodules can increase more than 200% when the number of effective nodules on the root system is reduced. The net result of this mechanism is that nodule tissue available to fix N is not a linear function of the number of nodules and the number of rhizobia in the soil. This mechanism may be a manifestation of legume adaptation to soil environments that vary in terms of the population of rhizobia available to support N_2 fixation.

Table 8-8 presents results from the WREN network and two additional sites in the Philippines. These experiments were carried out under high-management conditions to reduce the effects of environment between sites on legume response to inoculation. Results clearly indicate the effect that

Table 8-8. Influence of indigenous rhizobia on the response of legumes to inoculation with rhizobia tropical soils.†

No. rhizobia	Experiments	Relative inoculation response‡	Frequency of response	
			Inoculation	Nitrogen§
no. g^{-1} soil	no.		%	
0	14	252	71	86
1-10	16	94	92	71
1-100	13	14	70	70
100-1000	11	10	77	77
>1000	14	8	71	85

† Data from WREN: Puerto Rico (E. Schroeder); Ecuador (G. Bernal; C. Estevez); Morocco (A. Hilali); Taiwan (C.C. Young); Thailand (N. Boonkerd; P. Wadisirisuk); Zambia (R. Nyemba; V. Chinene); Egypt (H. Moawad); Hawaii (J. Thies, Ph.D. thesis, Univ. of Hawaii, 1990); Philippines (C. Escano, H. Layaoen, Y. Castroverde, and P. Singleton, 1990, unpublished data.)
‡ Legumes tested include: *Glycine max; Arachis hypogaea; Vigna unquiculata; Vigna radiata; Vicia faba; Leucaena leucocephala; Medicago sativa; Phaseolus lunatus; Phaseolus vulgaris; Trifolium repens; Vicia sativa; Lathyrus tingeatus; Trifolium subterraneum*. Response indicates inoculation increased seed yield or total crop nitrogen.
§ Nitrogen applications averaged >800 kg of N/(ha crop). Application rates recommended in the network were 150 kg of N every 2 wk during crop growth.

indigenous rhizobia have on the response to inoculation. Results from inoculation trials in the field are similar to results of the model presented in Fig. 8-3 and those of Thies et al. (1991b). The response to inoculation is large when there are few or no rhizobia in the soil. The response declines rapidly when there are more than 10 rhizobia g^{-1} of soil. The impact of indigenous rhizobia on the response to inoculation appears to be species neutral.

Results from six *Medicago sativa* experiments in the IBP inoculation trials measuring native rhizobia indicate an average relative increase in total crop N of 93% from inoculation when the indigenous rhizobial populations were <1 rhizobia g^{-1} of soil (Nutman, 1976). There is insufficient data from IBP trials conducted in soils with larger indigenous populations of rhizobia to determine whether the relationship between population size and response to inoculation is similar to the tropical legumes presented in Table 8-7.

Quantitative models predicting the response to inoculation in terms of measurable soil and environmental parameters will have to incorporate some measure of population density as a primary driving variable. The impact of population density on the response to inoculation clearly indicates the need to increase the precision and ease with which rhizobia in soil can be measured.

Effectiveness of Indigenous Rhizobia

Indigenous populations that are ineffective have been observed (Vincent, 1954; Jones et al., 1978). When populations of rhizobia are completely ineffective, a response to inoculation can be expected (Singleton & Tavares, 1986; Jones et al., 1978). Generally, however, populations of soil rhizobia are composed of numerous strains that vary in N_2-fixing capability with a particular host (Bergerson, 1970; Gibson et al., 1975; Singleton & Tavares, 1986; Trotman & Weaver, 1986; Vincent, 1954). It is difficult to quantify the relationship between population effectiveness of indigenous rhizobia and the response to legume inoculation when the population contains strains with a range of symbiotic effectiveness.

Brockwell et al. (1988) developed a method to measure the N_2-fixing effectiveness of rhizobial populations. Test plants are inoculated with a variable amount of soil to provide a predetermined number of rhizobia in the inoculum. Indication of population effectiveness is determined by growth of the test plant. Another measure of population effectiveness tested the effectiveness of many random rhizobial isolates from the soil and developed frequency distributions of the effectiveness of isolates within the population (Singleton & Tavares, 1986; Trotman & Weaver, 1986). Both methods have limitations. The first method requires accurate enumeration of soil rhizobia to avoid confounding the effectiveness measurement with rhizobial numbers. The second method is limited by the amount of effort required to test population by single isolates. Unless the indigenous population of rhizobia is completely ineffective, measures of population effectiveness have not been good indicators of the magnitude of the response to inoculation.

CONCLUSION

The myth that legumes do not respond to inoculation in the tropics is supported by general misconceptions about the promiscuity of tropical legumes and the ubiquity and abundance of tropical rhizobia. In reality, tropical legumes often exhibit some degree of specificity in their rhizobial requirements. In addition, many soils in the tropics do not contain sufficient indigenous rhizobial populations to meet the symbiotic potential of legume crops.

Limited data from previous site-specific inoculation trials have promoted the myth that tropical legumes do not respond to inoculation. While tropical legumes do not always respond to inoculation with sufficient magnitude to meet rigid statistical tests, the average response obtained from a range of tropical environments indicates that farmers will economically benefit from inoculation. Tropical legumes respond to inoculation in conditions fundamentally similar to those of temperate legumes. Specifically, the response to inoculation will be primarily controlled by the size and effectiveness of indigenous populations of rhizobia, and, to a lesser extent, by the yield potential of the crop and the availability of N from the soil.

Because the legume response to rhizobial inoculation involves a complex interaction between the crop, indigenous populations of rhizobia, and soil and environmental conditions, trials to determine the benefit from inoculation should account for these variables. These variables so profoundly affect the response to inoculation that site-specific results from inoculation trials are not reliable for making valid recommendations to farmers, government agencies, or private industry about the need to inoculate legumes.

The ultimate goal of continuing programs that involve inoculation trials should be to further quantify the complex interactions of variables determining the response to legume inoculation. Only then will it be possible to develop models based on soil and environmental test values that will predict the performance of legume inoculant in the field. Developing such capability is a prerequisite for determining where investment in inoculation technology is appropriate.

ACKNOWLEDGMENT

Research funding was provided in part by National Science Foundation Grant BSR-8516822, the Agency for International Development Cooperative Agreements DAN-4177-A-00-6035-00 (NifTAL Project), and DAN-146-G-SS-4081-00 (Indo-US Project/STI).

The authors wish to thank INLIT and WREN coordinators Robert Davis, Rudy Holthius, William Kerry, David McNeil, Joann Roskoski, Padma Somasegaran, Paul Singleton, and all INLIT and WREN collaborators. We also thank Susan Hiraoka for manuscript preparation, Janice Thies for graphics preparation, and Paul Woomer for technical assistance.

REFERENCES

Ahmad, M.H., A.R.J. Eaglesham, S. Hassauna, and B. Seaman. 1981. Examining the potential for inoculant use with cowpeas in West Africa. Trop. Agric. (Trinidad) 58:325–335.

Berg, R.K., R.M. Loynachan, R.M. Zablotowicz, and M.T. Lieberman. 1988. Nodule occupancy by introduced *Bradyrhizobium japonicum* in Iowa soils. Agron. J. 80:876–881.

Bergerson, F.J. 1970. Some Australian studies relating to the long term effects of the inoculation of legume seed. Plant Soil 32:727–736.

Bohlool, B.B., and E.L. Schmidt. 1973. Persistance and competition aspects of *Rhizobium japonicum* observed in soil by immunofluorescence microscopy. Soil Sci. Soc. Am. Proc. 37:561–564.

Boonkerd, N., D.F. Weber, and D.F. Bezdicek. 1978. Influence of *Rhizobium japonicum* strains and inoculation methods on soybeans grown in rhizobia-populated soil. Agron. J. 70:547–549.

Brockwell, J. 1977. Application of legume seed inoculant. p. 276–309. *In* R.W.F. Hardy and A.H. Gibson (ed.) A treatise on dinitrogen fixation. John Wiley and Sons, New York.

Brockwell, J., R.A. Holliday, and A. Pilka. 1988. Evaluation of the symbiotic nitrogen fixing potential of soils by direct microbial means. Plant Soil 108:163–170.

Burton, J.C. 1979. Rhizobium species. p. 29–57. *In* J. Peppler and D. Perlman (ed.) Microbial technology—Microbial processes. Academic Press, New York.

Cassman, K.G., D.N. Munns, and D.P. Beck. 1981a. Phosphorus nutrition of *Rhizobium japonicum*: Strain difference in phosphate storage and utilization. Soil Sci. Soc. Am. J. 45:517–520.

Cassman, K.G., D.N. Munns, and D.P. Beck. 1981b. Growth of *Rhizobium* strains at low concentrations of phosphate. Soil Sci. Soc. Am. J. 45:520–523.

Chen, W.X., G.H. Yan, and J.L. Li. 1988. Numerical taxonomic study of fast-growing soybean rhizobia and a proposal that *Rhizobium fredii* be assigned to *Sinorhizobium* gen. nov. Int. J. Syst. Bacteriol. 38:392–397.

Date, R.A. 1982. Assessment of the rhizobial status of the soil. p. 85–94. *In* J.M. Vincent (ed.) Nitrogen fixation in legumes. Academic Press, Sydney, Australia.

Davis, R.J., F.B. Cady, C.L. Wood, and C.P.Y. Chan. 1985. Design and analysis of an international experimental network: Legume inoculation trials in the NifTAL Project, the INLIT experience. HITAHR Res. Ser. 042. Coll. Trop. Agric. & Human Res., Univ. of Hawaii, Honolulu.

Diatloff, A., and S. Langford. 1975. Effective natural nodulation of peanuts in Queensland. Queens. J. Agric. Anim. Sci. 32:95-100.

Doku, E.V. 1969. Host specificity among five species in the cowpea cross-inoculation group. Plant Soil 30:126-128.

Dreyfus, B.L., J.L. Garcia, and M. Gillis. 1988. Characterization of *Azorhizobium caulinodans* gen. nov., a stem-nodulating, nitrogen fixing bacterium isolated from *Sesbania rostrata*. Int. J. Syst. Bacteriol. 38:89-98.

Evans, H.J., and L.E. Barber. 1977. Biological nitrogen fixation for food and fiber production. Science 197:332-339.

George, T. 1988. Growth and yield responses of *Glycine max* and *Phaseolus vulgaris* to mode of nitrogen nutrition and temperate changes with elevation. Ph.D. diss., Univ. of Hawaii, Honolulu.

George, T., P.W. Singleton, and B.B. Bohlool. 1988. Yield, soil nitrogen uptake, and nitrogen fixation by soybean from four maturity groups grown at three elevations. Agron. J. 80:563-567.

Gibson, A.H., B.L. Curnow, F.J. Bergerson, J. Brockwell, and A.C. Robinson. 1975. Studies of field populations of *Rhizobium*: Effectiveness of strains of *Rhizobium trifolii* associated with *Trifolium subterraneum* L. pastures in south-eastern Australia. Soil Biol. Biochem. 7:95-102.

Gibson, A.H., and J.E. Harper. 1985. Nitrate effect on nodulation of soybean by *Bradyrhizobium japonicum*. Crop Sci. 25:497-501.

Graham, P.H., and D.H. Hubbell. 1975. Legume-rhizobium relationships in tropical agriculture. p. 9-21. *In* E.C. Doll and G.O. Mott (ed.) Tropical forages in livestock production systems. ASA Spec. Publ. 24. ASA, Madison, WI.

Ham, G.E., V.B. Cardwell, and H.W. Johnson. 1971. Evaluation of *Rhizobium japonicum* inoculant in soils containing naturalized populations of rhizobia. Agron. J. 61:301-303.

Harris, S.C. 1979. Planning an international network of legume inoculation trails. NifTAL Project and Agency for Int. Develop., Paia, Hawaii.

Herridge, D.F., and J. Brockwell. 1988. Contributions of fixed nitrogen and soil nitrate to the nitrogen economy of irrigated soybean. Soil Biol. Biochem. 20:711-717.

Hinson, K. 1975. Nodulation responses from nitrogen applied to soybean half-root systems. Agron. J. 67:799-804.

Huang, C.Y., J.S. Boyer, and L.N. Vanderhoff. 1975. Limitation of acetylene reduction by photosynthesis in soybean having low water potentials. Plant Physiol. 56:228-232.

Imsande, J. 1986. Inhibition of nodule development in soybean by nitrate or reduced nitrogen. J. Exp. Bot. 37:348-355.

Johnson, H.W., U.M. Means, and C.R. Weber. 1965. Competition for nodule sites between strains of *Rhizobium japonicum*. Agron. J. 57:187-185.

Jones, M.B., J.C. Burton, and C.E. Vaughn. 1978. Role of inoculation in establishing subclover on California annual grasslands. Agron. J. 70:1081-1085.

Jordan, D.C. 1984. Family III Rhizobiaceae. CONN. 1938. p. 234-256. *In* N.R. Krieg (ed.) Bergey's manual of systematic bacteriology. Williams and Wilkins, Baltimore.

Keyser, H.H., and D.N. Munns. 1979. Tolerance of rhizobia to acidity, aluminum and phosphate. Soil Sci. Soc. Am. J. 43:519-523.

Lawson, K.A., Y.M. Barnet, and C.A. McGilchrist. 1987. Environmental factors influencing numbers of *Rhizobium leguminosarum* biovar *trifolii* and its bacteriophage in field soils. Appl. Environ. Microbiol. 53:1125-1131.

Meade, J., P. Higgins, and F. O'Gara. 1985. Studies of inoculation and competitiveness of a *Rhizobium leguminosarum* strain in soils containing indigenous rhizobia. Appl. Environ. Microbiol. 49:899-903.

Munns, D.N. 1977. Mineral nutrition and the legume symbiosis. *In* R.W.F. Hardy and A.H. Gibson (ed.) A treatise on dinitrogen fixation. John Wiley and Sons, New York.

Munns, D. 1968. Nodulation of *Medicago sativa* in solution: Acid sensitive steps. Plant Soil 28:129-146.

Nutman, P.S. 1976. Field experiments on nitrogen fixation by nodulated legumes. *In* P.S. Nutman (ed.) Symbiotic nitrogen fixation in plants. Cambridge Univ. Press, Cambridge.

Nyemba, R.C. 1986. The effect of *Rhizobium* strain, phosphorus applied, and inoculation rate on nodulation and yield of soybean [*Glycine max* L.) Merr. cv. 'Davis']. M.Sc. Diss. Univ. of Hawaii, Honolulu.

Parker, C.A., M.J. Trinick, and D.L. Chatel. 1977. Rhizobia as soil and rhizosphere inhabitants. p. 311-352. *In* R.W.F. Hardy and A.H. Gibson (ed.) A treatise on dinitrogen fixation. John Wiley and Sons, New York.

Ralston, E.J., and J. Ismande. 1983. Nodulation of hydroponically grown soybean plants and inhibition of nodule development by nitrate. J. Exp. Bot. 34:1371-1378.

Scholla, M.H., and G.H. Elkan. 1984. *Rhizobium fredii* sp. nov., a fast-growing species that effectively nodulates soybeans. Int. J. Syst. Bacteriol. 34:484-486.

Sellschop, J.P.F. 1962. Cowpeas, *Vigna unguiculata* (L.) Walp. Field Crop Abstr. 15:259-266.

Singleton, P.W., and B.B. Bohlool. 1983. Effect of salinity on the functional components of soybean-*Rhizobium japonicum* symbiosis. Crop Sci. 23:815-818.

Singleton, P.W., and K.R. Stockinger. 1983. Compensation against ineffective nodulation in soybean. Crop Sci. 23:69-72.

Singleton, P.W., H.M. AbdelMagid, and J.W. Tavares. 1985. Effect of phosphorus on the effectiveness of strains of *Rhizobium japonicum*. Soil Sci. Soc. Am. J. 49:613-616.

Singleton, P.W., and J.W. Tavares. 1986. Inoculation response of legumes in relation to the number and effectiveness of indigenous *Rhizobium* populations. Appl. Environ. Microbiol. 51:1013-1018.

Thies, J.E., P.W. Singleton, and B.B. Bohlool. 1991a. Influence of the size of indigenous rhizobial populations on establishment and symbiotic performance of introduced rhizobia on field-grown legumes. Appl. Environ. Microbiol. 57:19-28.

Thies, J.E., P.W. Singleton, and B.B. Bohlool. 1991b. Modeling symbiotic performance of introduced rhizobia in the field-based indices of indigenous population size and nitrogen status of the soil. Appl. Environ. Microbiol. 57:29-37.

Trotman, A.A., and R.W. Weaver. 1986. Number and effectiveness of cowpea rhizobia in soils of Guyana. Trop. Agric. 63:129-132.

Vincent, J.M. 1954. The root nodule bacteria of pasture legumes. Proc. Linn. Soc. N.S.W. 79:1-32.

Vincent, J.M. 1982. Role, needs, and potential of nodulated legumes. p. 263-284. *In* J.M. Vincent (ed.) Nitrogen fixation in legumes. Academic Press, Sydney, Australia.

Weaver, R.W., and L.R. Frederick. 1974a. Effect of inoculum rate on competitive nodulation of *Glycine max* L. Merrill. I. Greenhouse studies. Agron. J. 66:229-232.

Weaver, R.W., and L.R. Frederick. 1974b. Effect of inoculum rate on competitive nodulation of *Glycine max* L. Merrill. II. Field studies. Agron. J. 66:233-236.

Weaver, R.W., D.R. Morris, N. Boonkerd, and J. Sij. 1987. Populations of *Bradyrhizobium japonicum* in fields cropped with soybean-rice rotations. Soil Sci. Soc. Am. J. 51:90-92.

Weber, C.R. 1966. Nodulating and non-nodulating soybean isolines: II. Response to applied nitrogen and modified soil conditions. Agron. J. 58:46-49.

Woomer, P., P.W. Singleton, and B.B. Bohlool. 1988. Ecological indicators of native rhizobia in tropical soils. Appl. Environ. Microbiol. 54:1112-1116.

Yousef, A.N., A.S. Al-Nassiri, S.K. Al-Azawi, and N. Abdul-Hussain. 1987. Abundance of peanut rhizobia as affected by environmental conditions in Iraq. Soil Biol. Biochem. 19:319-396.

9 Impact of Soil Fauna on the Properties of Soils in the Humid Tropics[1]

P. Lavelle, E. Blanchart, and A. Martin
Ecole Normale Supérieure
Paris, France

A. V. Spain
1107 Ross River Road
Rasmussen, Queensland, Australia

S. Martin
Chargé de Mission au Ministère de l'Environnement
Neuilly-sur-Seine, France

The sustainability of soil fertility in agricultural systems of the humid tropics has recently become a major issue as a consequence of continued land degradation and the critical need to provide more food (FAO, 1981; Swaminathan, 1983). For socioeconomic, pedological, and ecological reasons, the development of sustainable high-input agriculture has proven to be slow and difficult and much effort needs to be directed towards the improvement of productivity in low-input agriculture.

In traditional low-input systems, sustainability is not achieved. Crop yields decrease rapidly following clearing and the land is generally abandoned after a few crops, providing a constant pressure to clear new land (e.g., Ayodele, 1986; Sanchez et al., 1983). Decreases in soil fertility are due not only to nutrient loss exported as crops (e.g., Sanchez et al., 1989). Other factors accelerating fertility losses include:

1. The erosion of exposed soil leading to the loss of the fine particles rich in organic matter and nutrients.
2. Overmineralization of soil organic matter (SOM) and plant residues due to the overheating of bare soil.

---

[1] Contribution from Laboratoire d'Ecologie de l'Ecole Normale Superieure de Paris (UA CNRS no. 258). The authors express appreciation to Ministère de l'Environnement (SRETIE, Relations Internationales) for financial assistance.

Copyright © 1992 Soil Science Society of America and American Society of Agronomy, 677 S. Segoe Rd., Madison, WI 53711, USA. *Myths and Science of Soils of the Tropics.* SSSA Special Publication no. 29.

3. Nutrient leaching by high water fluxes and a lack of synchrony between nutrient release and plant demand.
4. Deterioration of soil structure consequent on the decreased activity of soil macroorganisms (macrofauna and roots).

The situation in adjacent natural systems provides a marked contrast. Primary productivity levels of several tens of megagrams of dry mass ha^{-1} are generally sustained, even in apparently poor infertile soils (e.g., Herrera et al., 1978; Bernhardt-Reversat et al., 1979) and the differences are not readily explained solely by the export of crops from the exploited systems. Nearly closed nutrient cycles ensure an optimal conservation of mineral fertility whereas soil structure is maintained through diverse processes, particularly intense biological activity. It is, therefore, critical to understand the nature and relative importance of the biological processes that maintain fertility, to use them in cropping systems and thus improve the sustainability of their use (Swift, 1984).

This chapter first discusses the relative importance of biological processes, particularly those mediated by the soil fauna, on the maintenance of fertility in the soils of the humid tropics. Because of the relatively recent recognition of the soil fauna as constituting a major controlling influence on the productivity of tropical ecosystems, only two important myths have emerged. The first of these suggests that the earthworms (*Lumbricidae*) are important in the soils of the temperate areas while the termites are the dominant animals of the tropics. The second states that earthworms are the product rather than the cause of high fertility. In refuting these myths, the composition of macroinvertebrate communities and the relative importance of their components in the major ecosystems of the humid tropics are considered. Finally, the effects of the two dominant macrofaunal groups, the earthworms and the termites on SOM dynamics and soil structure are detailed.

In this chapter, we confine discussion largely to the saprophagous organisms that process dead plant materials and SOM or exert a major influence on the soil processes that guide the functioning of its systems and evolution. We, therefore, largely omit discussion of the ants despite their undoubted influence on soil communities since most operate at the higher trophic levels in nature. It is only the leaf-cutter ant (tribe Attini) that derive most of its energy from cellulose and thus act as active herbivores.

BIOLOGICAL SYSTEMS OF REGULATION AND SOIL FERTILITY

Soil fertility has two major components:
1. Nutrient supply to plants which in turn depends on: (i) the intensity and spatial and temporal pattern of nutrient release through the mineralization of decomposing organic matter and (ii) the ability of soil to retain a store of cations in exchangeable form on the surfaces of colloids (clay minerals and SOM). Decomposition is thus shown to be a key process for soil fertility through its effects on both mineralization and humification.

2. Soil physical structure, which controls the void space and thus the amounts of water and oxygen stored in the soil and the rate of their supply to roots. Decomposition and physical structure are both determined by a suite of hierarchically organized factors that operate at different temporal and spatial scales (Anderson et al., 1989; Lavelle et al., 1991) (Fig. 9-1). At the highest level of the hierarchy, the climate (i.e., precipitation and temperature regimes) generally accounts for much of the large-scale variation in decomposition rates and parent material weathering (Heal et al., 1981; Pedro, 1983).

At the next level down, parent material characteristics strongly influence soil texture, nutrient status, and both the abundance and nature of the clay minerals. They thus determine both decomposition rates and soil physical structure. At an even lower level, the quality of organic inputs greatly affects decomposition rates through the different ways in which chemicals inhibit microbial activity (e.g., the inhibition of N release from resistant phenol-protein complexes and situations where the C/nutrient ratios are imbalanced or lignin contents are high).

Finally, at the lowest level of resolution, integrated groups of organisms formed into biological systems of regulation (Lavelle, 1984) directly mediate decomposition through their digestive activities and the production of metabolites, feces, and casts with distinctive physical and chemical properties. Biological systems of regulation are composed of macroorganisms (in-

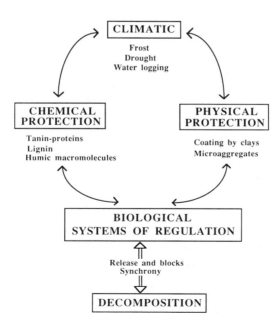

Fig. 9-1. Hierarchy of factors that determine microbial activity and eventually decomposition rates.

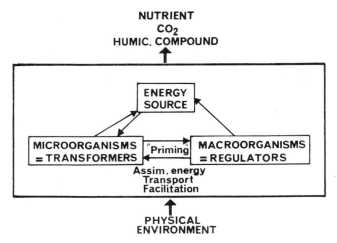

Fig. 9-2. General organization of biological systems of regulations in soils (Lavelle, 1984).

vertebrates and roots) and microorganisms (mainly fungi and bacteria) whose activity is directly controlled by that of the macroorganisms (Fig. 9-2).

Microbial communities are able to effect almost any kind of chemical transformation, and certainly may even change their individual metabolic functions in response to altered environmental conditions (e.g., Swift, 1976; Visser, 1985). However, microorganisms have extremely limited capacities for movement and may thus remain for long periods in inactive resting stages because they are physically separated from their food resources or lack suitable microenvironmental conditions. This is especially true of bacteria, whereas fungi are relatively more mobile and less dependent on specific, localized environmental conditions.

Calculated turnover times for microbial biomass give unexpectedly high estimates of about 1 to 2 yr in temperate soils, and more than 0.5 yr in tropical soils (Jenkinson & Ladd, 1981; Chaussod et al., 1986) while generation times may be < 20 h under optimum laboratory conditions (Clarholm & Rosswall, 1981). Much of the microbial population thus spends a large proportion of the time in dormant stages with low metabolic activity awaiting favorable conditions. During this period, their size progressively diminishes to the point beyond which reactivation is no longer possible.

Reactivation of dormant microorganisms and propagules normally results from the activity of macroorganisms that brings them into contact with new nutritional substrates (by mixing the soil or moving into it) and provides them with water and readily assimilable organic matter (in the form of root exudates or earthworm cutaneous and intestinal mucus). Microorganisms initially utilize these resources to reach the high metabolic levels at which they are able to use more complex organic substrates (Barois & Lavelle, 1986; Coleman et al., 1983). Such "priming effects" (Jenkinson, 1966) have been measured in the rhizosphere in response to exudate production (Billes et al., 1986) and in the gut of the endogeic earthworm (drilosphere) as a

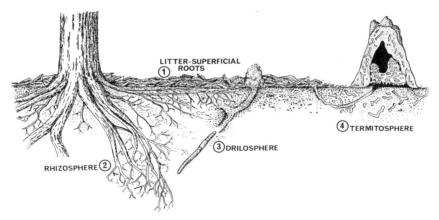

Fig. 9-3. The four principal biological systems of regulation. (After Lavelle, 1984.)

response to intestinal mucus (Barois & Lavelle, 1986; Martin et al., 1987). Similar effects may be triggered in the guts and galleries of termites (termitosphere) and the litter system when water-soluble organic matter leached by rain contacts the dormant microorganisms of the subsoil.

There are four principal biological systems of regulation (Fig. 9-3).

1. The first system is the litter-superficial roots system, the main function of which is a direct cycling of nutrients from the leaf litter to surface roots, thus preventing losses. In that system, litter is the energy source. The microflora is dominated by fungi and macroorganisms that comprise a rich and diverse community of litter-feeding arthropods and Oligochaetes (epigeic earthworms and Enchytraeidae). Surface roots growing among the decomposing leaves act as a sink for the nutrients released.
2. The second system is the rhizosphere. It is a focus for the active mutualistic interactions between roots, microflora, and an associate microfauna (Protozoa and Nematodes). The system is activated by the production of root exudates.
3. The third is the drilosphere, which comprises the geophagous earthworm populations and the soil that they influence (i.e., intestinal contents, castings, and galleries).
4. The fourth is the termitosphere, which is defined as termites and the whole soil and microflora that they influence through their activities. Microbial communities are diverse and occasionally highly specific (e.g., gut symbionts and *Termitomyces* spp. that are the characteristic fungi cultivated in the fungus gardens of macrotermitine termites).

In the soils of the humid tropics, the critical level in the hierarchy of the factors determining soil fertility is variable. Climatic controls become ineffective when temperatures are constantly elevated and moisture limits activity for only short periods. The effect of clay minerals (kaolinite and oxyhydroxides) as regulators of decomposition and the basis of the soil structure is limited in many soils due to their frequently low reactivities and occasional low abundance. Consequently, decomposition and fertility processes are likely to be regulated at the lowest level (i.e., that of resource quality and biological systems of regulation). Thus, soil invertebrates are likely to be the major regulators of decomposition in many tropical situations. This is particularly true of macroinvertebrates whose effects on soil microorganisms may be dramatic due to the relatively large scales of space and time at which they operate. Macroinvertebrates also have major effects on soil structure by forming macropores, digging galleries, and egesting the casts and pellets that are often the basic components of well-developed macroaggregate structures.

SOIL MACROINVERTEBRATE COMMUNITIES IN THE HUMID TROPICS

Classification of Soil Invertebrates

Soil organisms face three major constraints. They feed on relatively low-quality resources, move in a compact and amphibious environment, and must be preadapted to survive occasionally unfavorable temperature and moisture conditions.

Three main strategies have been developed to cope with these constraints.

1. *Microbionts* (i.e., microorganisms and the microfauna) are small (<0.2 mm) and live in water-filled pores. Microorganisms have a considerable ability to digest low-quality substrates and may survive unfavorable temperature and moisture conditions by sheltering in soil micropores. Nonetheless, they have a limited capacity to move and no access to the physically protected organic matter present within soil microaggregates. The most important groups of microfauna are the Nematoda and Protozoa.

2. *Macrofauna* (>1 cm) form a marked contrast in that their enzymatic capacities are generally limited and they have little ability to withstand harsh environmental conditions. However, they can move through the soil by digging galleries and burrows. They break the physical protection of soil organic matter during the process of ingesting and mixing the soil in their gut. One part of the macrofauna does not enter the soil but remains in the leaf litter since it possesses no adaptations for digging. This section includes a range of large insects (Coleoptera and Diptera larvae) and other Arthropoda (e.g., Isopoda, Myriapoda, and Arachnida). The main soil-dwelling components of the macrofauna are the earthworms, termites, and ants, some of which may also live in the litter and feed on it.

3. The *mesofauna* (0.2–1.0 mm) has an intermediate strategy in that the members of this group move through existing soil cracks and natural channels, but are unable to move through the undisturbed soil. They are usually confined to the surface litter and have a well-developed resistance to drought and extreme temperatures. This group comprises microarthropods (i.e., Col-

lembola, Acari, and several other groups) and small whitish worms of the Oligochaete family Enchytraeidae.

Only the macrofauna will be considered here as the abundance of smaller animals and their effects on fertility are still poorly documented and appear to operate at very small scales. Their activities appear, in some cases, to be controlled by those of higher organisms (e.g., nematodes and protozoa in the micro-food chain of the rhizosphere, or Collembola and Acari in litter systems dominated by the activities of the larger arthropod or earthworm).

The macrofauna, in its turn, may be divided into three main functional groups, depending on the kind of food they use (surface litter or underground litter or SOM) and their location in the profile (litter, subsoil, or burrows and nests):

1. *Epigeics* live in the litter; they feed on decomposing litter, live microorganisms (mainly fungi), or invertebrates. They have no effect on the soil structure as they are unable to dig into the soil. Their main effect is the fragmentation, comminution, and partial digestion of litter. They comprise a very diverse community of saprophagous and predatory arthropods (e.g., microarthropods, myriapods, isopods, and arachnids), some pigmented earthworms and a few other groups of invertebrates (e.g., gastropods);

2. *Endogeics* live in the soil. They mainly feed on soil organic matter and dead roots. Live roots seem to be a poor resource as they are seldom consumed by endogeics (see e.g., Lavelle, 1983; Lavelle et al., 1989b). Endogeics have been divided into oligo-, meso-, and polyhumics according to how they ingest: (i) soil of the deep horizons with a low organic content; (ii) the bulk soil of the upper 10 to 25 cm; and (iii) soil of the upper horizon enriched in organic matter by preferentially feeding in organic-rich pockets of soil (e.g., rhizosphere) or ingesting fine particles with a high organic content rather than coarse sand particles. Endogeics are active diggers that may greatly enhance soil aggregation. They mainly comprise earthworms, humivorous termites, and a few rhizophagous arthropods.

3. *Anecics* feed in the litter but live in the soil in subterranean nests or galleries, or in epigeic nests. Their main effect is to export litter from the litter system to other systems in which the time courses and pathways of decomposition are different. They also affect soil physical characteristics (e.g., water infiltration and aeration by digging galleries). This category includes the large-pigmented earthworms and a large proportion of termites.

Macrofauna Communities

In the humid tropics, the nature of the diverse macroinvertebrate communities depends largely on vegetation type (Fig. 9-4). Forests generally have highly diverse communities in which the epigeic litter arthropods and anecic termites are dominant in population density terms, although endogeic and anecic earthworms may be major components of biomass. In the more humid savannas and pastures, the endogeic earthworms are always a dominant component of biomass. In natural savannas, however, termites are frequent-

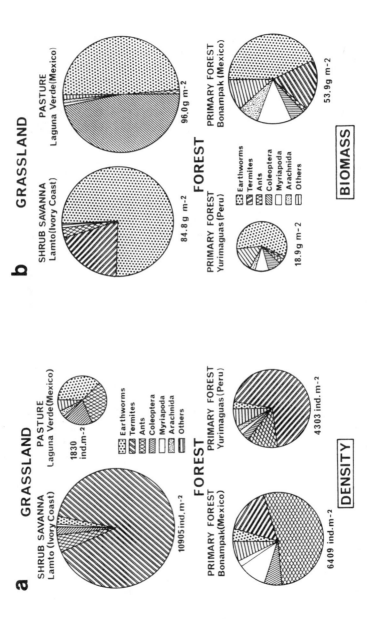

Fig. 9-4. Density (a) and biomass (b) structure of macrofaunal communities in grasslands and forests of the humid tropics (Lavelle et al., 1981; Lavelle, 1983; Lavelle & Kohlmann, 1984; Lavelle & Pashanasi, 1989).

IMPACT OF SOIL FAUNA ON SOIL PROPERTIES

ly dominant or important and significant populations of ants or litter or root-feeding arthropods may be present.

The termites (Isoptera) are a group of social exopterygote insects allied to the cockroaches (Blattodea). Species diversity decreases markedly from the lower to the higher latitudes of 45 °N and 45 °S. Species diversity also decreases sharply with increasing elevation (Wood, 1979, 1988; Collins, 1983).

The occurence of the termite and distribution of its feeding habits depends mostly on biogeographic factors and climatic patterns. In America, fungus-cultivating termites (Termitidae, Macrotermitinae) are not represented and termite abundance is generally limited as compared to equivalent climatic areas of Africa and Australia. Australia, while not possessing fungus-cultivating termites, has a rich and diverse termite fauna.

Earthworm communities also exhibit similar large-scale variations. This is especially true in forests where geographic variation is noticeable. It has been found, however, that beyond biogeographical differences the abundance and distribution of populations into ecological categories are ultimately determined by rain. "Dry" forests have relatively low populations, dominated by epigeic species while, in moderately wet areas (1800–2500 mm of annual rainfall) endogeic species often predominate and biomass is at a maximum. In wetter areas, where high leaching has impoverished the soils, earthworm biomass tends to be lower and endogeics are partly replaced by anecic and epigeic populations. Finally, in areas receiving more than 4000 to 5000 mm annually, earthworm populations are generally small (Fragoso & Lavelle, 1990).

Earthworms and termites are by far the most active components of the soil fauna with respect to their influences on soil structure and SOM dynamics. Earthworms are especially active in the upper 10 to 20 cm of soil and, when abundant, may exert a day-to-day regulation of microbial activity with immediate as well as medium-term effects (months to years). Termites are active throughout and even below the profile and their effects on soil structure and SOM dynamics are manifested partly on a short-term basis and partly in the long term (years to decades). Another major difference between termites and earthworms is that the activities of the latter are rather uniformly distributed on a horizontal dimension whereas termite activities are concentrated in their nests and galleries.

EARTHWORM EFFECTS: THE DRILOSPHERE SYSTEM

The earthworm communities of the humid tropics are largely dominated by endogeic populations that use SOM as a nutritional resource. As a result, the overall effects of earthworm communities on soils differ substantially between the humid tropics and the temperate areas that are dominated by epigeic and anecic populations feeding on leaf litter (Lavelle, 1983). This section examines the effects of the endogeic earthworm on the soil physical structure and SOM dynamics in the humid tropics.

Effect of the Earthworm on Soil Physical Structure

Drilosphere Structures

The endogeic earthworm has a major effect on soil structure by promoting macroaggregation (i.e., the combination of soil particles into stable compound structures). Its effect on water infiltration and the vertical mixing of soil horizons may also be significant, even though it does not build burrow systems and only leaves evenly distributed macropores in the soil it colonizes.

Individuals may ingest daily from 5 to 30 times their own weight of soil, depending on the species and soil conditions (Lavelle, 1975, 1988). As a result, field populations with biomasses of 200 to 1000 kg fresh wt ha^{-1} may annually ingest from a few hundred to 1250 Mg of dry soil ha^{-1} (Lavelle, 1978). Most of this soil is egested into the subsoil. Only a small proportion is egested at the soil surface as casts. Two main kinds of casts may be distinguished, that is: (i) globular casts of high stability, comprised of coalescent round or flattened subunits and (ii) granular casts comprised of an accumulation of small, fragile, fine-textured pellets, with little structural stability.

An annual production of surface casts of 14 to 15 Mg ha^{-1} has been measured in tropical Costa Rican pastures colonized by high populations of the pantropical endogeic earthworm *Pontoscolex corethrurus* (Fraile, 1989). Even higher figures of 25 to 30 Mg ha^{-1} have been recorded from the moist savannas of the Ivory Coast and the Cameroons (Kollmannsperger, 1956; Lavelle, 1978). However, at Lamto this only represents 1.7 to 3.5% of the 800 to 1250 Mg of dry soil that the endogeic earthworm ingests annually. Interestingly, this proportion varies seasonally probably as a response to soil compaction, and is at a maximum when the overall soil ingestion is minimum (Fig. 9-5).

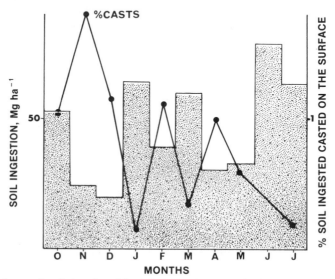

Fig. 9-5. Seasonal variations in soil ingestion (blocks) and surface cast (solid line) production by earthworms at Lamto (Ivory Coast). (After Lavelle, 1978.)

Soil Aggregation

In humid tropical soils with high endogeic earthworm activity, casts deposited in the subsoil are the component units of stable macroaggregate structures. Such macroaggregates may comprise 50 to 60% of the soil (Blanchart, 1990) (Fig. 9-6). Experiments conducted on savanna soils at Lamto (Ivory Coast) have demonstrated that macroaggregate structures may form in the presence of the endogeic earthworm in surprisingly short periods. In 33 d, the activity of 5 g fresh weight of *Millsonia anomala* in 6800 g of soil resulted in the formation of 2883 g of aggregates (i.e., 42.4% of the soil in the experimental container), as compared to 1075 g (15.8%) in a treatment with the perennial grass *Panicum maximum* Jacq. and 906 g (13.3%) in a control treatment with neither plants nor earthworm (Blanchart et al., 1990). In such a short period, only 1815 g (26.7%) of aggregates might have been produced as earthworm casts according to earlier results (Lavelle, 1975). This indicates that factors other than the simple production of globular casts contribute to aggregation. The difference (i.e., 15.7%) might be due to the proliferation of fungal hyphae that may have linked together soil particles (Tisdall & Oades, 1982).

When earthworms are excluded from a soil within which they had formed a macroaggregate structure, this structure remains stable for a long time (at least several years) due to the stabilization of aggregates with time. Alternation of dry and wet periods seems to particularly favor this aggregate stabilization. The introduction of earthworms producing small granular casts, however, leads to the progressive destruction of the structure as large aggregates are split into much smaller and more fragile ones (Fig. 9-7). It is hypothesized that, in natural ecosystems such as the savannas at Lamto (Ivory Coast), excessive soil aggregation that may negatively affect plant growth (e.g., Rose & Wood, 1980; Spain et al., 1990) is prevented by the activities of the small earthworms that maintain soil macroaggregation at less than a maximum value of approximately 60% of the soil as macroaggregates larger than 2 mm (Blanchart, 1990).

Water Infiltration

Earthworms may also greatly affect water infiltration. In West African soils, Casenave and Valentin (1988) reported a fivefold increase of infiltration in the presence of earthworms and termites as compared with soils where they were absent.

Regulation of Soil Organic Matter Dynamics

In soils of the humid tropics, endogeic earthworm activities significantly affect SOM dynamics both in the short term (acceleration of nutrient release) and long term (physical protection of SOM in casts). Such effects are a direct consequence of the mutualistic relationship that links these worms to the soil microflora in a common exploitation of SOM resources (Barois & Lavelle, 1986; Martin et al., 1987). These effects are generally significant

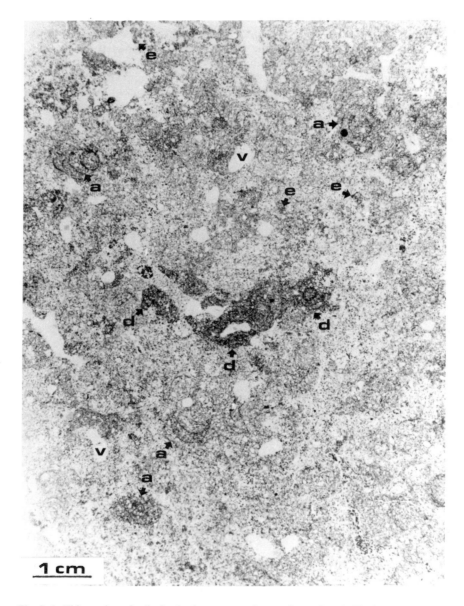

Fig. 9-6. Thin section of soil of a shrub savanna at Lamto (Ivory Coast). The large earthworm fills its galleries with large impacted casts (a). These casts have a thin external layer rich in clay minerals and organic matter; the innerpart has a higher porosity. Small filiform Eudrillidae produce compact casts with a high organic content (e). Large casts with a fine texture (d) are also apparent. Voids (v) are numerous, due to compaction of soil in casts and deposition of part of the ingested soil as surface casts.

IMPACT OF SOIL FAUNA ON SOIL PROPERTIES

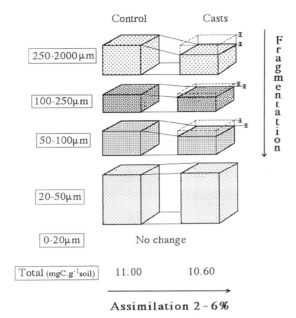

Fig. 9-7. Changes in abundance and particle size structure of soil organic matter of a savanna soil (Lamto, Ivory Coast) after digestion by the endogeic earthworm *Millsonia anomala* (Martin, 1989).

at the scale of the whole soil profile as a consequence of high ingestion rates. At Lamto, the 1000 Mg ha^{-1} ingested by endogeic earthworms, on average may contain 15 Mg SOM (i.e., one-third of the total SOM accumulated in a grass savanna soil), and 60% of the SOM in the 10 upper cm where earthworm activity is concentrated (Lavelle, 1978). Other estimates made in Mexican (C. Fragoso, 1990, unpublished data) and Costa Rican pastures (Fraile, 1989) give ingestion figures of about 350 to 500 Mg ha^{-1} and 17.5 to 25 Mg of SOM.

Short-Term Effects

Assimilation of Soil Organic Matter. The assimilation of SOM by earthworms ranges from 2 to 9% in the African species *Millsonia anomala* to a maximum of 18% in the pantropical *Pontoscolex corethrurus* (Lavelle, 1978; Martin et al., 1990; Barois et al., 1987).

By using natural ^{13}C labelling of SOM particle-size fractions, it has been shown that *M. anomala* digests equally the coarsest organic fractions (assumed to be young and relatively assimilable) and the finest ones (assumed to be old and resistant) (Martin et al., 1991b).

During digestion, a clear change occurs in the distribution of SOM among the particle-size fractions: in the casts of the African endogeic *M. anomala*, the relative abundance of organic matter associated with coarse fractions decreased sharply ($-25%$ in the 250–2000 µm fraction) while that

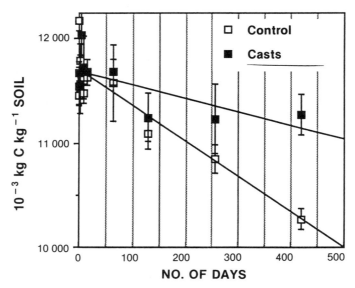

Fig. 9-8. Variation in total C content of casts of the endogeic earthworm *Millsonia anomala* and a 2 mm-sieved control soil in long-term laboratory incubations (savannah sandy alfisol, field capacity, 27 °C).

in the finer fractions tended to increase. This difference mainly resulted from comminution (fragmentation without digestion) of the largest fragments, since all fractions were assimilated at the same rate (Fig. 9-8).

Release of nutrients. Transit of SOM through the earthworm gut results in the release of significant amounts of assimilable nutrients. Fresh casts of *P. corethrurus* fed on the soil of the upper 10 cm of a 15-yr-old secondary forest at Yurimaguas (Peru) contained up to 133 10^{-3} kg kg^{-1} mineral N divided into 111 mg ammonium (NH_4^+) (i.e., 18.2 times the value in uningested soil) and 22.4 mg nitrate (NO_3^-) (1.12 times the control) (Lavelle, Melendez, Schaefer & Pashanasi, In preparation). Comparable measurements made on the casts of *M. anomala* fed on a savanna soil at Lamto (Ivory Coast) only contained 24.5 10^{-3} kg kg^{-1} NH_4^+ and only 0.5 10^{-3} kg kg^{-1} NO_3^- (P. Lavelle and S. Martin, 1990, unpublished data). In this case, it represented a net mineralization rate of 3% and an overall annual release by natural field populations of 13 to 18 kg ha^{-1}.

Ammonium concentration rapidly decreased in the casts. In the savanna soil of Lamto, nitrification was negligible and NH_4^+ concentration decreased below control levels after 8 d. In a forest soil of Peruvian Amazonia, NH_4^+ concentration in casts of *P. corethrurus* decreased much more slowly with time. After 16 d, NH_4 concentration in casts was still 2.5 times the control value. Part of it had undergone nitrification and the resulting NO_3^- concentration was 71.1 mg kg^{-1}, that is, 3.55 times the initial control value. This decrease of mineral N in casts was probably due to microbial reorganization since no leaching was permitted to occur, no plants were avail-

Table 9-1. Available forms of P (water-extractable and exchangeable) in casts of *P. corethrurus* of different ages from soils with a low (Lamto) and high (Laguna Verde) P retention capacity (Lopez-Hernandez, Lavelle, and Fardeau, 1991, unpublished data) (means accompanied by the same letter are not significantly different; $P < 0.05$).

	Water extractable P	Exchangeable P
	10^{-6} kg kg^{-1} soil	
Lamto (sandy alfisol)		
Control soil	0.060 +/− 0.005 a	2.00 +/− 0.11 a
Casts 12 h	0.158 +/− 0.005 b	6.00 +/− 1.78 b
Casts 96 h	0.499 +/− 0.10 c	15.89 +/− 1.78 c
Laguna Verde (vertisol)		
Control soil	0.026	15.00 +/− 0.78 b
Casts 12 h	0.021	10.11 +/− 1.78 a
Casts 24 h	0.053	43.33 +/− 1.11 c
Casts 96 h	0.032	11.44 +/− 1.11 a

able to act as a sink and denitrification would have been unlikely (P. Lavelle et al., 1990, unpublished data).

Among other nutrients critical for plant growth, P availability seems to be affected by endogeic earthworm activities. In casts of *P. corethrurus*, exchangeable and water extractable inorganic P increase with time. Maximum values of, respectively, two to eight times and three to eight times the control have been measured after 1 to 4 d (D. Lopez-Hernandez et al., 1990, unpublished data) (Table 9-1).

Long-term Effects

It seems probable that in soils with high endogeic populations, little, if any, resistant organic matter would accumulate in the upper 20 cm of the profile where earthworm activity concentrates since no organic pool seems resistant to decomposition in their gut. In soils where such an endogeic drilosphere effect predominates, the turnover of SOM is thus expected to be extremely rapid. This has been observed in the Lamto savanna where endogeic earthworm populations are abundant and active: a natural ^{13}C labelling of SOM was effected due to a rapid afforestation of the savanna after protection from fire. The half-life time of SOM was estimated at, respectively, about 8 and 19 yr for the coarse 50 to 2000 μm "labile" and 2 to 50 yr "resistant" particle-size fractions (Martin et al., 1990). Despite the clear acceleration of turnover due to the mutualistic digestive system of the earthworms (in association with the free-living soil bacteria), a system of SOM protection in casts of *M. anomala* subsequent further to digestion, has been described (Martin, 1989). Mineralization of SOM decreases sharply with time in the casts and stabilizes at a minimum after a few days. As a result, SOM content of casts soon overtook values measured in a control noningested 2-mm sieved soil, incubated under similar conditions (Fig. 9-8) and after 1 yr, the total amount of SOM mineralized in casts was 40% less than in the control.

Earthworm Activities and Plant Growth

The effect of the endogeic earthworm on SOM dynamics is thus that of a "release-and-block" system with phases of extremely intensive mineralization, during digestion, and further inhibition of mineralization, probably due to physical protection of SOM in the compact structure of the casts. One peculiarity of this system is that it may well function in synchrony with plant demand for nutrients: in a microcosm experiment associating the endogeic worm (*M. anomala*) and a plant (*Panicum maximum* Jacq.) ^{15}N-labelled worms rapidly eliminated their ^{15}N as NH_4 in their casts and part of this ^{15}N was subsequently transferred to the plant (Spain et al., 1991, in press).

In pot experiments, improvements in plant growth in the presence of the endogeic earthworm have been reported. In the above-mentioned experiment, the growth of *P. maximum* Jacq. increased with increasing earthworm biomass up to a maximum value of 2.8 times the control after 70 d. Nitrogen concentration was increased by, respectively, 29 to 116% in leaves and 88 to 260% in roots. Phosphorus content was increased in roots (+58–152%) while no significant change was observed in the leaves.

At Yurimaguas (Peruvian Amazonia), the effect of *P. corethrurus* on the growth of seedlings of three different tree species was assessed during a 120-d growth period (Pashanasi et al., 1991, unpublished data). *Bixa orellana* grew three times higher in the presence of earthworms than in their absence; *Eugenia stipitata* grew 1.3 to 1.7 times higher while no significant effect was observed with *Bactris gasipaes*. The different responses of tree seedlings were probably due to variation in the abundances and distributions of the root systems that would allow a much better use of nutrients released by the earthworm in the case of *B. orellana* (dense, deep-rooting system) than in *B. gasipaes* (low density of coarse superficial roots). Dramatic effects on N mineralization and microbial biomass were observed.

Recent studies (Redell & Spain, in preparation) show that endogeic, and other earthworms, strongly influence the distribution of mycorrhizal propagules. This particularly applies to the spores of vesicular-arbuscular mycorrhizae that are concentrated in their casts, probably through preferential feeding in the rhizosphere (Spain et al., 1990b). Part of the growth simulation previously considered may thus result from an earlier and greater level of mycorrhizal infection of the host plant. In addition, casts of these earthworms possess higher levels of some plant-available nutrient elements and their placement in areas of low soil strength make them preferential root growth and infection pathways.

TERMITE EFFECTS: THE TERMITOSPHERE SYSTEM

The termites are not a diverse group. Taxonomically, they are divided into the "higher" termites of the family Termitidae while the "lower" termites comprise the remaining five families. Grassé (1986) estimated that more than 2000 species of termites are currently known, the majority of which are found in the tropics.

Where common, termites play important ecological roles. Apart from the few species that attack living plants, almost all termites feed on dead organic matter, breaking it down in a complete way and sequestering SOM and plant nutrients for considerable periods in their nests, mounds, galleries, and other structures. Erosion of aboveground structures constantly returns these nutrients to the soil surface and underground structures also decay allowing their contained nutrient elements to be again taken up by plants. This leads to a nutrient patchiness in the landscape that preserves the ecological diversity. In many infertile environments, the only place where plants requiring a high nutrient status may grow is on or adjacent to termitaria. The extensive underground gallery systems created, greatly improve soil aeration and water movement. Termites may also help to restart biological nutrient cycling in degraded ecosystems through the same mechanism.

Termite colonies are divided into several castes permitting a division of labor. Reproduction is normally confined to a single pair of individuals, the king and the queen. Workers forage for food and build the mounds while highly specialized soldiers defend the colonies against predators and the inroads of termites from surrounding colonies. Termite colonies reproduce by periodically releasing reproductives that mate and then may establish new colonies separate from those within which they were reared. However, new colonies may also establish from secondary reproductives where part of the original colony becomes isolated from its parent, or the original queen dies.

Termites have their greatest influence in the ecosystems of the tropics and subtropics, and it is there that they also have their greatest effects on soils. The effects of termites on soil morphology and processes are significant at scales ranging from the catena (e.g., Wielemaker, 1984), to the microaggregate (e.g., Garnier-Sillam et al., 1985) and even the clay mineral level (Boyer, 1982).

Termites and Soil Textural Relationships

Termite distribution and activity are both affected by, and affect soil texture. Mound-building termites are largely excluded from vertisols on the savanna lands of northern Australia, and elsewhere, due to the instability induced by the expansive smectitic clays present (e.g., Holt & Coventry, 1982). A lack of clay may also limit mound-building termites; the wood-feeding species *Coptotermes acinaciformis* is excluded from deep sandy soils in Western Australia, apparently for this reason (Lee & Wood, 1971).

Termites substantially affect surface processes by transporting soil from various depths in the profile to the surface, where it is built into epigeal mounds and other structures. Surface structures erode continuously leading to a redistribution of soil materials in the landscape. Because termites generally select the smaller particles from within the profile, materials brought to the surface are commonly finer in texture and may have a different clay mineral composition than those of the original surface (Boyer, 1982). Thus, abstraction of finer particles and their surface deposition lead to the gradual build up of a stone-free surface soil (e.g., Williams, 1968). Net annual increments in the thickness of such soils have been estimated to range from 0.01 to 0.10 mm yr^{-1} (Lee & Wood, 1971). This is an important pedological and ecological process in the large areas of the tropics and subtropics where the termite fauna is dominated by species that form epigeal mounds. Beneath the surface, changes in texture, porosity, and the distribution of organic and mineral matter are some of the many consequences of their presence.

Effects of Termites on Soil Structure

Termites have profound effects on soil structure through their mound-building activities, aggregate formation, and the creation of systems of subterranean galleries and storage chambers. Although the so-called subterranean termites do not build mounds, they may have considerable influence on soil structure. Nonetheless, epigeal mounds are highly visible structures of substantial importance in terms of pedogenesis and plant growth, both directly and in terms of their erosion products. While a range of termite-derived structures are known from mounds and gallery walls, less is known of their contribution to soil aggregation away from the immediate environs of the mounds or of the role of subterranean termites in this regard. Termite galleries act as a network of horizontal and vertical macropores influencing bulk density, aeration, and water infiltration and permitting movement of materials both upwards and downwards within the soil (e.g., Wielemaker, 1984). Movement of these soil materials through the galleries is mediated both by the activities of the termites and physical processes.

Mounds

Mounds may be abundant; Aloni and Soyer (1987) report populations of more than 5000 mounds ha^{-1} from Zaire, principally of the humivorous termite *Cubitermes*. The aboveground parts of termite mounds may weigh some hundreds of tons per hectare and occupy considerable proportions of the soil surface, depending on their type. In tropical Australia, mounds may number up to 1100 ha^{-1} representing approximately 62 t ha^{-1} of soil and covering 1.7% of the sampled area (Wood & Lee, 1971). However, the median mass for the 58 sites recorded in Spain et al. (1983) is only 20 Mg ha^{-1} (interquartile range 15-25) and the median area covered by their bases is only 0.8% of the sampled areas (interquartile range 0.5-1.1). Similar estimates from Africa range up to 10% of the area sampled. Although Meyer (1960) recorded the bases of mounds at one site as covering an exceptional 33% of the soil surface, representing an estimated dry mass of 2400 t ha^{-1}. Where populations of mounds are high, they are frequently overdispersed (Spain et al., 1986) suggesting that termite populations may tend to fully use the normal organic resources available to them in the landscape. This is reinforced by the denudation and dietary shifts that occur during drought years (Lepage, 1981).

The epigeal mounds built by termites are usually fine-textured, cemented, hard and massive structures that shed almost all the water that impinges on them. Through the energy of rainfall impact and runoff, a steady erosion of mound surfaces occurs even where they still contain healthy populations (Bonell et al., 1986). It is possible that this may limit the geographic distribution of some termite species whose mounds lack specific adaptations for water shedding. There must be a substantial energy cost to termite colonies attributable to erosional losses during mound building and repair; the soil materials recently added to epigeal termitaria are particularly susceptible to erosion, at least until hardened by curing or internal repacking processes.

After the colony has died, or the mound has been broken up, the resistant mound materials erode slowly and may continue to modify local drainage patterns for considerable periods. Where mounds are common, an area equivalent to the entire soil surface will be covered by their bases over quite short geological times. For example, the median area of the bases of the mounds recorded from the northern Australian sites given above is 0.8% of the land surface area (Spain et al., 1983). Given that this figure remains constant over time, an area equivalent to the whole surface will have been covered by the bases of termite mounds in approximately 125 generations of mounds. If the turnover time of the mound materials is assumed to be of the order of 30 to 50 yr, this represents a period of 3730 to 6250 yr, a short period in terms of soil formation.

Simple calculations of this type patently underestimate the effect of termite mounds on the soil. Many incipient colonies build small mounds but fail through the effects of predation, competition, or other factors. In addition, the soil materials brought to the surface to form runways and cover food supplies has only occasionally been quantified. The few data available suggest that this may be in the order of 1 Mg ha^{-1} yr^{-1} (see, e.g., Bagine, 1984).

The turnover times of mound soil materials are poorly known, but are substantially longer than the life spans of the colonies that construct the mounds. Individual mounds of *Nasutitermes triodiae* in northern Australia are known to have existed for approximately 100 yr (Watson, 1972) and medium-sized mounds from a range of species and localities are recorded as having existed for 30 to 115 yr (Grassé, 1984). For the soil materials of a population of small mounds (principally humivores of the genus *Cubitermes*) from Zaire, Aloni and Soyer (1987) estimated a turnover time of approximately 10 yr. In addition to erosion by raindrop impact and runoff water, mounds are also subject to damage by termitophilous vertebrates and cattle.

Aggregate Formation

Sleeman and Brewer (1972) have described a wide range of structures from termite mounds. More recently, Kooyman and Onck (1987) also used micromorphological concepts in categorizing termite-derived structures in soils distant from mounds. Apart from fragments of mound materials, storage structures and gallery walls, few aggregates have been directly attributed to termite activity. Garnier-Sillam et al. (1985), however, have recently described organo-mineral microaggregate structures from the feces of four species of termites of different ecological strategies and Eschenbrenner (1986) has remarked on the similarity of aggregates in termite-inhabited soils to those from the mounds. It is apparent that much progress remains to be made in this area of study.

The effects of termites on soil structure depend on the ecological strategies of the local termite population. Higher percentages of water-stable aggregates have been recorded by several authors from termite-affected soils. This is illustrated by the work of Garnier-Sillam et al. (1988) who report contrasting effects on soil structure by two sympatric species of termites from the Congo. In the situation studied, a humivorous termite, *Thoracotermes macrothorax* builds mounds and other structures with a high organic matter content and a relatively low level of clay enrichment. In contrast, the mounds of the fungus-cultivating species *Macrotermes mulleri* and their associated surface soils, were highly clay-enriched and had low organic matter contents. Both the mound wall materials of *M. mulleri* and the soil from the A1 horizon associated with them were markedly less permeable to water and less stable than the control surface soils while with *T. macrothorax*, the opposite pertained. These authors noted a strong correlation between the organic matter content of the soil and stability. A surface-sealing effect was also associated with dispersion of the clay-enriched materials from the mounds of *M. mulleri*. This has been noted elsewhere, particularly in conjunction with degrading termitaria.

Galleries

The importance of termite galleries in influencing the physical properties of soils has been recognized for some time in locations where populations are large (Lee & Wood, 1971), although few quantitative data are available. In an experimental study in a desert ecosystem, Elkins et al. (1986) reported higher bulk densities, lower infiltration rates, and higher bedloads in runoff waters from areas where subterranean termites had been eliminated.

It is apparent from the studies conducted so far that the nature, depth distribution, and extent of such galleries varies markedly with the species involved. For example, those of the humivorous termites are usually more superficial than those of the Macrotermitinae. Termite galleries vary in shape from circular to narrowly elliptical, and in size, from < 1 mm to more than 20 mm in cross section (Grassé, 1984). Gallery lengths of up to 7.5 km ha^{-1} (Darlington, 1982) have been recorded from soils associated with the mounds of *Macrotermes michaelseni* although Wood (1988) felt that this figure could perhaps have been doubled to account for those of the subterranean termite species present. In addition, the equivalent of 90 000 storage chambers ha^{-1} were also present in the study area. Assuming that these structures were largely confined to the top 20 cm of the soil and applying Wood's estimate, it may be calculated that termite galleries and related structures occupied approximately 0.4% of the soil volume of this zone.

Termites and Soil Organic Matter

As previously stated, termites feed basically on cellulosic materials. However, the feeding habits of soil-dwelling termites cannot be classified in the same way as those of earthworms or other soil animals. This results from the fact that while these termites make their nests and live in soil or mounds, most species derive their nutrition from above or close to the soil surface. In addition, termites do not limit their activities to any one horizon. Most species are thus surface or near-surface feeding saprovores; some species harvest the dead foliage of grasses, seeds, the feces of higher animals and other materials from the soil surface. A wide range of species attack dead wood and perform a useful role in recycling these materials in nature, although they frequently damage man's wooden constructions. The species of one further group, the humivores, feed on organic matter within the soil. Several species also feed on living plants both above and below ground and some are serious pests of crops and other plants.

The ecological impact of termite feeding may be large. Lepage (1972) reports that termites in low-lying parts of sahelian grasslands of Sénégal may consume up to 49% of annual herbage production, although the overall figure for the area was 5.4%. In further studies of semiarid pastures in Kenya, Lepage (1981) estimated that termites and grazing mammals had similar impacts both consuming approximately 1 Mg ha^{-1} yr^{-1}. Lee and Wood (1971) have reported that *Nasutitermes exitiosus* consumed 16% of the total estimated fall of leaves and sticks or 4.9% of the total annual litter fall in an Australian dry sclerophyll forest. Similar figures for Malaysian rainforests range from 1% for a site prone to flooding to 16% for a relatively dry forest whose termite fauna was dominated by Macrotermitinae (Collins, 1983).

In terms of the impact of termite grazing on soil, Lepage (1981) reported that African *Macrotermes* spp. may, in drought years, shift their diet from normal litter-feeding habits to feeding on standing herbaceous vegetation and, with increasing severity of drought, to feeding on roots. This destabilizes the soil surface and renders it readily susceptible to erosion and degradation. In sites where the Macrotermitinae are the dominant termites, it appears that a higher proportion of annual litter fall may be recycled through termites and this has been ascribed to the high efficiency of the symbiotic relationship with species of the fungus genus *Termitomyces*, described below.

Under some circumstances, a relationship between termite presence and soil organic matter concentrations has been demonstrated. This was shown by Parker et al. (1982) who found increased organic matter in desert soils following removal of the termites present. Similarly, the absence of fire and an effective termite fauna from the vertisol studied by Moore et al. (1967) in subtropical Australia, were considered to be the causes of the large-standing crop biomass of aboveground dead organic matter. Moore et al. (1967) estimated that there were 75 Mg ha^{-1} of litter together with 58 Mg ha^{-1} of standing dead material on the site in addition to elevated levels of soil organic matter and N.

Termite Digestive Processes

A feature of many termite foodstuffs is their relative recalcitrance to breakdown; most are low in N and materials normally considered as readily assimilable. Nonetheless, assimilation rates in the termites are high and published values range between 54 and 93% of the food eaten (Wood, 1978). Termite foodstuffs are normally rich in lignin and other resistant plant compounds. This has led to the apparently universal associations with microorganisms which, with their wider spectrum of enzymatic capacities, contribute much to the breakdown of the resistant material eventually assimilated.

Digestive mechanisms differ between the more highly evolved Termitidae and the remaining or lower termite families. In the lower termites, digestion of resistant materials appears to be carried out largely by the obligately anaerobic protozoans that inhabit a specialized region of the hind gut, while in the higher termites, these protozoans are absent or ineffective. In addition, in the lower termites, a wide range of bacteria play diverse but, as yet ill-defined nutritional roles. In the higher termites, the animal's own enzymes are perhaps more effective in carbohydrate breakdown. Although there is also a prominant fermentation in the hind gut with acetic acid being the major end product of cellulose breakdown in all termites (Breznak, 1984). In addition, acquired enzymes play an important role in cellulose digestion in at least some Macrotermitinae (Martin, 1984) while members of this subfamily derive much of their nutrition through the degradational activities of the basidiomycete fungus *Termitomyces* cultivated in their nests (Thomas, 1987a,b).

Lignin appears to be degraded by termites although the mechanisms are as yet unclear, since the known methods of degradation are aerobic (Kirk & Farrell, 1987). Butler and Buckerfield (1979) and Cookson (1987) demonstrated the evolution of labelled CO_2 from live termites that had been fed lignin labelled with ^{14}C in various parts of its structure. Cookson (1987) found that both higher and lower termites had some capacity to degrade lignins although their capacities to do so differed between lignins from different sources.

Nitrogen-fixing organisms have been found in the gut of termites and efficient N conserving mechanisms are known from the Isoptera (Breznak, 1984). These include behavioral traits such as cannibalism, necrophily, and the ability of at least some termites to reabsorb uric acid from the hind gut.

Organic Matter in Termite-derived Structures

Carbon levels in epigeal termite structures are normally higher than those of the soils from which they are formed, because of the salivary or fecal materials that are frequently used as adhesives. One such salivary adhesive from the mound wall of the wood-feeding species *Coptotermes acinaciformis* has been identified as a glycoprotein (Gillman et al., 1972). In structures formed mainly from soil materials, the degree of organic matter enrichment depends primarily on whether feces are included. Thus, the C content of the mounds of the Macrotermitinae is lower than that of the humivores because the former do not include feces in the mounds. The C/N ratio is higher in many termite-derived materials than in the surrounding surface soils although the organic matter of the humivorous species is an obvious exception to this. Lee and Wood (1971) showed that this ratio was higher in mound materials from wood-feeding species than in grass-feeding species.

The centers of the mounds of certain wood- and grass-feeding termites consist of carton, a mixture of excreta, some inorganic materials and fragments of undigested, comminuted plant material. Organic matter in carton is high and that in the mounds of *Coptotermes acinaciformis*, a wood-feeding termite from northern Australia, ranged from 83 to 93% loss-on-ignition (Lee & Wood, 1971).

Soils beneath and adjacent to termitaria are also modified, often by physical processes such as the erosion of materials from mounds. In tropical Australia, the mounds of litter and grass-feeding termites (*Amitermes* spp.) are surrounded by an erosional pediment that has C levels similar to the mound materials, or intermediate between these and soils further from the mounds. While plants only occasionally grow on the mound surfaces because of their hardness and low water contents, the increased fertility of the pediments is evidenced in the field by the dominance of the annual grasses and forbs. These contrast with the perennial grasses and sedges found on the less-fertile soils further from the mounds (Arshad, 1982; Spain & McIvor, 1988). The enhanced glasshouse growth of test plants on materials from these mounds (Okello-Oloya & Spain, 1986) and from soils beneath and adjacent to the mounds (Spain & Okello-Oloya, 1985) further evidences their relatively higher fertility than surface soils distant from the mounds.

There have been few detailed studies of the nature of the organic matter of termite mounds. Lee and Wood (1971) presented information on the levels of organic matter and selected organic and inorganic components in mound materials and the carton of several Australian wood and grass-feeding termites. Other studies have concentrated on the mounds of the Macrotermitidae. Arshad et al. (1988) studied the organic matter from the mounds of two African Macrotermitinae (*Macrotermes michaelseni* and *M. herus*). Generally, humic acids from soils were shown to have higher molecular weights and less aromatic materials than those from the mounds. In moving from the outer wall of the mounds of the *Macrotermes* spp. to the nursery to the royal chamber, the concentration of CO_2H groups increased and that of the carbohydrate and other O-substituted carbon decreased. Some intersite variation in the composition of the organic matter of the *Macrotermes* mounds was also noted. Carton from the Australian mounds studied was shown to be high in lignin and humic acids, with higher lignin, total carbohydrate and N concentrations and C/N ratios in the wood-feeding species than in the grass-feeding species.

The fungal comb of *M. michaelseni* has been studied by Arshad and Schnitzer (1987) and Arshad et al. (1988). These authors found that its ash content was approximately 14%. The approximate composition of the organic matter was 40% carbohydrate (all of which occur as polysaccharides which on hydrolysis were shown to be dominated by glucose), 10% protineaceous material (all as peptides or longer-chain structures). Appreciable aliphatic and aromatic plus phenolic materials were also noted. Using fractionation techniques commonly used for soil materials, the base-insoluble humin (40% of organic matter) was found to be mostly carbohydrate while the humic acid (40%) and fulvic acid (20%) components contained most of the aromatic material.

The $\delta^{13}C$ values (proportional distribution of the ^{13}C and ^{12}C carbon isotopes) of organic matter in the soil materials from termitaria in the savannas of Lamto (Côte d'Ivoire) are up to 5.5 $d^{13}C$ units more negative than those of the surrounding surface soil (A.V. Spain and M. Lepage, 1988, unpublished data). The more negative values of the mound materials may be due to those of the lignin-enriched humified organic matter of the termite feces. Alternatively, some of the organic matter of the mounds may be derived from the woody (C_3) vegetation that grows on the surface of the mounds in addition to the (C_4) grasses.

DISCUSSION

The two major myths addressed in the introduction have been shown to be demonstrably inaccurate. Both earthworms and termites clearly influence tropical ecosystems, both natural and man-modified. Their distribution patterns are closely related to rainfall, vegetation, edaphic, and biogeographic factors. Earthworms are confined to areas with an annual rainfall >800 mm on the average. Termites exist in much drier climatic conditions. They become predominant in semiarid and arid environments.

The second myth that assumes that earthworm abundance might be a consequence rather than the cause of soil fertility is refuted through an understanding of the processes relating production processes to soil structure and the maintenance of fertility.

Our present knowledge of the role of soil macroinvertebrates in the humid tropics suggests that they may be efficient agents in the maintenance of soil fertility. Apart from termites and earthworms, other soil faunal components are also important. For example, arthropods of the litter system that accelerate the transformation and incorporation of litter into the soil and the microfauna that represents an intermediate level of regulation between macro- and microorganisms (i.e., bacteria and fungi) are important. At the higher trophic levels, the ant exerts considerable predation pressure on the soil meso- and macrofauna and are sometimes important pedogenetic agents (Holldobler & Wilson, 1990).

It has been observed several times that certain agricultural practices are detrimental to soil faunal communities of natural ecosystems (Critchley et al., 1979; Lavelle & Pashanasi, 1989). Cultures of maize leave little of the original fauna and no adaptable alternative fauna is generally present. Systems associating trees (e.g., peach-palm *Bactris gasipaes* or *Hevea*) with a plant cover of legumes may conserve a significant part of the original fauna whereas pastures may occasionally be colonized by peregrine endogeic earthworm species (like, e.g., *Pontoscolex corethrurus*) that can build up large populations with biomasses of up to 1 to 4 Mg fresh wt. ha^{-1}.

These observations are still too scarce. The possibility of maintaining or improving soil fertility by manipulating the activities of soil fauna needs to be explored. This field of research, however, is regarded as having considerable potential (see e.g., Swift, 1984; Lavelle et al., 1989). Preliminary short-term experiments (on short scales of time and space) have demonstrated that considerable improvements in plant production and the maintenance of fertility may be gained through a proper management of this resource.

A better knowledge is required of the biology of a wide range of species adaptable to the relatively harsh ecological conditions of, for example, rice fields, yucca plantations, or the environment of dry forest-like tree plantation crops (e.g., cocoa and tea). Species adaptable to such an environment may exist, but in distinct geographical areas. Earthworm species adapted to live in a maize field of the Amazon are not likely to be found in the nearby rainforest, but rather in savannas of South America and Africa whose soils have temperature and moisture regimes comparable to those of a cultivated field.

It is also important to have information on the dynamics of soil colonization by adaptable populations. Adaptable species may exist in an area, but may be impeded from colonizing because they are separated from the field by such natural barriers as a small stream or a few rows of trees.

Large-scale experiments are required to assess the effect of introducing adaptable species (or removing some, such as termite species feeding on live materials) over significant areas (about 1 ha) and during several successive crops, to ascertain whether the improvements observed over short periods in small experimental containers, are still maintained under conditions much closer to those of normal agricultural practice.

Finally, more attention should be paid to the often fertile soil accumulated in structures and mounds built by litter and grass-harvesting termites. The exploitation of this soil as a fertilizer, on a regular basis is practiced in a limited number of cases (Swift et al., 1989). Research should be made to evaluate the feasibility of such practices on a larger scale.

REFERENCES

Aloni, K., and J. Soyer. 1987. Cycles des matériaux de construction des termitières d'humivores en savane au Shaba méridional (Zaire). Rev. Zool. Afr. 101:329-357.

Anderson, J.M., and P. Flanagan. 1989. Biological processes regulating organic matter dynamics in tropical soils. p. 97-125. *In* D.C. Coleman et al. (ed.) Dynamics of soil organic matter in tropical ecosystems. NifTAL Project, Univ. of Hawaii, Honolulu.

Arshad, M.A. 1982. Influence of the termite *Macrotermes michaelseni* (Sjostedt) on soil fertility and vegetation in a semi-arid savannah ecosystem. Agro-Ecosystems 8:47-58.

Arshad, M.A., and M. Schnitzer. 1987. The chemistry of a termite fungus comb. Plant Soil 98:247-256.

Arshad, M.A., M. Schnitzer, and C.M. Preston. 1988. Characterization of humic acids from termite mounds and surrounding soils, Kenya. Geoderma 422:213–225.

Ayodele, O.J. 1986. Effect of continuous maize-cropping on yield, organic carbon mineralization and phosphorus supply of savannah soils in western Nigeria. Biol. Fertil. Soil. 2:151–155.

Bagine, R.K.N. 1984. Soil translocation by termites of the genus *Odontotermes* (Holmgren) (Isoptera: Macrotermitinae) in an arid area of northern Kenya. Oecologia (Berlin) 64:263–266.

Barois, I. 1987. Interactions entre les Vers de terre (Oligochaeta) tropicaux géophages et la microflore pour l'exploitation de la matière organique du sol. Travaux des Chercheurs de Lamto (RCI), no. 7. ENS, Paris.

Barois, I., and P. Lavelle. 1986. Changes in respiration rate and some physico-chemical properties of a tropical soil during transit through *Pontoscolex corethrurus* (Glossoscolecidae, Oligochaeta). Soil Biol. Biochem. 18:539–541.

Barois, I., B. Verdier, P. Kaiser, A. Mariotti, P. Rangel, and P. Lavelle. 1987. Influence of the tropical earthworm *Pontoscolex corethrurus* (Glossoscolecidae, Oligochaeta) on the fixation and mineralization of nitrogen. p. 151–159. *In* A.M. Bonvicini and P. Omodeo (ed.) On earthworms. Mucchi Editore, Modena.

Bernhardt-Reversat, F., C. Huttel, and G. Lemée. 1979. Structure et fonctionnement des écosystèmes de la forêt pluvieuse sempervirente de la Côte d'Ivoire. p. 605–625. *In* Écosystèmes forestiers tropicaux. Recherches sur les Ressources Naturelles XIV. UNESCO, Paris.

Billes, G.N., N. Gandais-Riollet, and P. Bottner. 1986. Effet d'une culture de graminées sur la décomposition d'une litière végétale, marquée au ^{14}C et ^{15}N, dans le sol, en conditions controlées. Acta Oecol., Oecol. Plant. 7(3):273–286.

Blanchart, E. 1990. Rôle des vers de terre dans la formation et la conservation de la structure des sols de la savane de Lamto (Côte d'Ivoire). Thèse Univ. Rennes I.

Blanchart, E., P. Lavelle, and A. Spain. 1990. Effects of biomass and size of *Millsonia anomala* (Oligochaeta, Acanthodrilidae) on particle aggregation in a tropical soil in presence of *Panicum maximum* (Gramineae). Biol. Fertil. Soils 9:5.

Bonell, M., R.J. Coventry, and J.A. Holt. 1986. Erosion of termite mounds under natural rainfall in semi-arid tropical northeast Australia. Catena 13:11–28.

Boyer, P. 1982. Quelques aspects de l'action des termites du sol sur les argiles. Clay Miner. 17:453–462.

Breznak, J.A. 1984. Biochemical aspects of symbiosis between termites and their intestinal microbiota. p. 173–203. *In* J.M. Anderson et al. (ed.) Invertebrate-microbial interactions. Cambridge Univ. Press, Cambridge.

Butler, J.H.A., and J.C. Buckerfield. 1979. Digestion of lignin by termites. Soil Biol. Biochem. 11:507–511.

Casenave, A., and C. Valentin. 1988. Les états de surface de la zone sahélienne. Leur influence sur l'infiltration. Rapport CEE-ORSTOM, ORSTOM, Bondy, France.

Chaussod, R., B. Nicolardot, G. Catroux, and C. Chrétien. 1986. Relations entre les caractéristiques physico-chimiques et microbiologiques de quelques sols cultivés. Sci. Sol. 2:213–226.

Clarholm, M., and T. Rosswall. 1980. Biomass and turnover of bacteria in a forest soil and a peat. Soil Biol. Biochem. 12:49–57.

Coleman, D.C., C.P.P. Read, and C.V. Cole. 1983. Biological strategies of nutrient cycling in soil systems. Adv. Ecol. Res. 13:1–55.

Collins, N.M. 1983. Termite populations and their role in litter removal in Malaysian rainforests. p. 313–325. *In* S.L. Sutton et al. (ed.) Tropical rainforest: Ecology and management. Blackwell Sci. Publ., Oxford.

Cookson, L.J. 1987. ^{14}C-lignin degradation by three Australian termite species. Wood Sci. Technol. 21:11–25.

Critchley, B.R., A.G. Cook. U. Critchley, T.J. Perfect, A. Russell-Smith, and R. Yeadon. 1979. Effects of bush clearing and soil cultivation on the invertebrate fauna of a forest soil in the humid tropics. Pedobiologia 19:425–438.

Darlington, J.P.E.C. 1982. The underground passages and storage pits used in foraging by a nest of the termite *Macrotermes michaelseni* in Kajiado, Kenya. J. Zool. 198:237–247.

Elkins, N.Z., G.V. Sabol, T.J. Ward, and W.G. Whitford. 1986. The influence of subterranean termites on the hydrological characteristics of a Chihuahuan desert ecosystem. Oecologia (Berlin) 68:521–528.

Eschenbrenner, V. 1986. Contribution des termites à la microagrégation des sols. Cah. ORSTOM, sér. Pedol. 22:397–408.

Food and Agriculture Organization. 1981. Agriculture: Horizon 2000. Vol. 23. Développement Economique et Social. FAO, Rome.

Fragoso, C., and P. Lavelle. 1991. Earthworm communities of tropical rainforests. Soil Biol. Biochem. (in press).

Fraile, J.M. 1989. Poblaciones de Lombrices de tierra (Oligochaeta: Annelida) en una pastura de *Cynodon pletostachvus* (Pasto estrella) asociada con aboles de *Eythina poeppigiana* (Poro), una pastura asociada con arboles de *Cordia alliodora* (Laurel), una pastura sin arboles y vegetacion a libre crecimiento, en el CTIE, Turrialba, Costa Rica. Tesis Doctoral, Univ. de Costa Rica.

Garnier-Sillam, E., G. Villemin, F. Toutain, and J. Renoux. 1985. Formation de micro-agrégats organo-mineraux dans les feces de termites. C.R. Acad. Sci. Ser. 3 301:213–218.

Garnier-Sillam, E., F. Toutain, and J. Renoux. 1988. Comparaison de l'influence de deux termitières (humivore et champignoniste) sur la stabilité structurale des sols forestiers tropicaux. Pedobiologia 32:89–97.

Gillman, L.R., M.K. Jeffries, and G.N. Richards. 1972. Non-soil constituents of termite (*Coptotermes acinaciformis*) mounds. Aust. J. Biol. Sci. 25:1005–1013.

Grassé, P.P. 1984. Termitologia. Vol. 2. Masson, Paris.

Grassé, P.P. 1986. Termitologia. Vol. 3. Masson, Paris.

Herrera, R.A., C.F. Jordan, H. Klinge, and E. Medina. 1978. Amazon ecosystems: Their structure and functioning with particular emphasis on nutrients. Interciencia 3:233–240.

Holldobler, B., and E.O. Wilson. 1990. The ants. Harvard Univ. Press, Harvard.

Holt, J.A. 1987. Carbon mineralization in semi-arid northeastern Australia: The role of termites. J. Trop. Ecol. 3:255–263.

Holt, J.A., and R.J. Coventry. 1982. Occurrence of termites (Isoptera) on cracking clay soils in northeastern Queensland. J. Aust. Entomol. Soc. 21:135–136.

Jenkinson, D.S. 1966. The priming action. p. 198–207. *In* The use of isotopes in soil organic matter studies. J. Appl. Radiat. Isotopes, Spec. Suppl., Pergamon, Oxford.

Josens, G. 1983. The soil fauna of tropical savannas. III. The termites. p. 498–513. *In* F. Bourlière (ed.) Tropical savannas. Elsevier, New York.

Kirk, T.K., and R.T. Farrell. 1987. Enzymatic "combustion": The microbial degradation of lignin. Ann. Rev. Microbiol. 41:465–505.

Kollmansperger, F. 1956. Lumbricidae of humid and arid regions and their effect on soil fertility. p. 293–297. *In* 6th Congr. Int. Sci. Sol.

Kooyman, C., and R.F.M. Onck. 1987. Distribution of termite (Isoptera) species in southwestern Kenya in relation to land use and the morphology of their galleries. Biol. Fert. Soils 3:69–73.

Lavelle, P. 1975. Consommation annuelle de terre par une population naturelle de vers de terre (*Millsonia anomala*) Omodeo, Acanthodrilidae-Oligochètes) dans la savane de Lamto (Côte d'Ivoire). Rev. Ecol. Biol. Sol. 12:11–24.

Lavelle, P. 1978. Les Vers de terre des savanes de Lamto (Côte d'Ivoire): Peuplements, populations et fonctions dans l'écosystème. Publications du Laboratoire de Zoologie, Ecole Normale Supérieure, Paris.

Lavelle, P. 1983. The soil fauna of tropical savannas. I. The community structure. II. Earthworm communities. p. 477–484 and 485–497. *In* F. Bourlière (ed.) Tropical savannas. Elsevier, New York.

Lavelle, P. 1984. The soil system in the humid tropics. Biol. Int. 9:2–17.

Lavelle, P. 1988. Earthworm activities and the soil system. Biol. Fert. Soils 6:237–251.

Lavelle, P., I. Barois, A. Martin, Z. Zaidi, and R. Schaefer. 1989a. Management of earthworm populations in agroecosystems: A possible way to maintain soil quality? p. 109–122. *In* M. Clarholm and L. Bergström (ed.) Ecology of arable land. Developments in Plant and Soil Sciences 39. Kluwer Academic Publ., Dordrecht, Holland.

Lavelle, P., R. Schaefer, and Z. Zaidi. 1989b. Soil ingestion and growth in *Millsonia anomala*, a tropical earthworm, as influenced by the quality of the organic matter ingested. Pedobiologia 33:379–388.

Lavelle, P., E. Blanchart, A. Martin, S. Martin, A.V. Spain, F. Toutain, I. Barois, and R. Schaefer. 1991. A hierarchical model for decomposition in terrestrial ecosystems: Application to soils of the humid tropics. Biotropica (in press).

Lavelle, P., and B. Kohlmann. 1984. Etude quantitative de la macrofaune du sol dans une forêt tropicale mexicaine (Bonampak, Chiapas). Pedobiologia 27:377–393.

Lavelle, P., and B. Pashanasi. 1989. Soil macrofauna and land management in Peruvian Amazonia (Yurimaguas, Loreto). Pedobiologia 33:283–291.

Lee, K.E., and T.G. Wood. 1971. Termites and soils. Academic Press, New York.

Lepage, M. 1972. Recherches écologiques sur une savane sahélienne du Ferlo septentrional, Sénégal: données préliminaires sur l'écologie des termites. La Terre Vie 26:383–409.

Lepage, M. 1981. L'impact des populations récoltantes de *Macrotermes michaelseni* (Sjostedt) (Isoptera: Macrotermitinae) dans un écosystème semi-aride (Kajiado, Kenya). II. La nourriture récoltée, comparaison avec les grands herbivores. Insect. Soc. 28:309–319.

Martin, A. 1989. Effet des vers de terre tropicaux géophages sur la dynamique de la matière organique du sol dans les savanes humides. Thèse doctorat. Univ. Paris-Sud.

Martin, A., J. Cortez, I. Barois, and P. Lavelle. 1987. Les mucus intestinaux de ver de terre, moteurs de leurs interactions avec la microflore. Rev. Ecol. Biol. Sol 24:549–558.

Martin, A., A. Mariotti, J. Balesdent, P. Lavelle, and R. Vuattoux. 1990. Estimate of organic matter turnover rate in a savanna soil by ^{13}C natural abundance measurements. Soil Biol. Biochem. 22(4):517–523.

Martin, A., A. Mariotti, J. Balesdent, and P. Lavelle. 1991. Estimate of soil organic matter assimilation by a geophagous tropical earthworm based on ^{13}C natural abundance. Ecology (in press).

Martin, M.M. 1984. The role of ingested enzymes in the digestive processes of insects. p. 155–172. *In* J.M. Anderson et al. (ed.) Invertebrate-microbial interactions. Cambridge Univ. Press, Cambridge.

Meyer, J.A. 1960. Résultats agronomiques d'un essai de nivellement des termitières réalisé dans la cuvette centrale Congolaise. Bull. Agric. Congo Belge 51:1047–1059.

Moore, A.W., J.S. Russell, and J.E. Coldrake. 1967. Dry matter and nutrient content of a subtropical semiarid forest of *Acacia harpophylla* F. Muell. (Brigalow). Aust. J. Bot. 15:11–24.

Okello-Oloya, T., and A.V. Spain. 1986. Comparative growth of two pasture plants from northeastern Australia on the mound materials of grass and litter-feeding termites (Isoptera: Termitidae) and on their associated surface soils. Rev. Ecol. Biol. Sol 23:381–392.

Okello-Oloya, T., A.V. Spain, and R.D. John. 1985. Selected chemical characteristics of the mounds of two species of Amitermes (Isoptera, Termitinae) and their adjacent surface soils from northeastern Australia. Rev. Ecol. Biol. Sol 22:291–311.

Parker, L.W., H.G. Fowler, G. Ettershank, and W.G. Whitford. 1982. The effects of subterranean termite removal on desert soil nitrogen and ephemeral flora. J. Arid Environ. 5:53–59.

Pedro, G. 1983. Structuring of some basic pedological processes. Geoderma 31:289–299.

Pullan, A. 1979. Termite hills in Africa: Their characteristics and evolution. Catena 6:267–291.

Rose, C.J., and A.W. Wood. 1980. Some environmental factors affecting earthworm populations and sweet potato production in the Tari Basin, Papua, New Guinea. Agric. J. 31:1–13.

Sanchez, P.A., D.E. Bandy, and J.H. Villachica. 1983. Soil fertility dynamics after clearing a tropical rainforest in Peru. Soil Sci. Am. J. 47:1171–1178.

Sanchez, P.A., C.A. Palm, L.T. Szott, E. Cuevas, and R. Lal. 1989. Organic input management in tropical agroecosystems. p. 125–152. *In* D.C. Coleman et al. (ed.) Dynamics of soil organic matter in tropical ecosystems. NifTAL Project, Univ. of Hawaii, Honolulu.

Sleeman, J.R., and R. Brewer. 1972. Microstructures of some Australian termite nests. Pedobiologia 12:347–373.

Spain, A.V., P. Lavelle, and A. Mariotti. 1992. Preliminary studies of the effects of some tropical earthworms on plant growth. Soil Biol. Biochem. (in press).

Spain, A.V., and T. Okello-Oloya. 1985. Variation in the growth of two tropical pasture plants on soils associated with the termitaria of *Amitermes laurensis* (Isoptera: Termitinae) p. 141–145. *In* R.B. Chapman (ed.) Proc. 4th Australasian Conf. on Grassland Invertebrate Ecology. Caxton Press, Christchurch, NZ.

Spain, A.V., T. Okello-Oloya, and A.J. Brown. 1983. The abundances, above-ground masses and basal areas of termite mounds at six locations in tropical north-eastern Australia. Rev. Ecol. Biol. Sol. 20:547–566.

Spain, A.V., D.F. Sinclair, and P. Diggle. 1986. Spatial distribution of the mounds of harvester and forage termites (Isoptera: Termitidae) at four locations in tropical north-eastern Australia. Oecol. Gener. 7:335–352.

Spain, A.V., and J.G. McIvor. 1988. The nature of herbaceous vegetation associated with termitaria in north-eastern Australia. J. Ecol. 76:181–191.

Spain, A.V., P.G. Saffigna, and A.W. Wood. 1990a. Tissue carbon sources for *Pontoscolex corethrurus* (Oligochaeta; Glossoscolecidae) in a sugarcane ecosystem. Soil Biol. Biochem. 22:703–706.

Swaminathan, M.S. 1983. Our greatest challenge: Feeding the hungry world. p. 25–31. *In* G. Bixlet and L.W. Shemilt (ed.) Chemistry and the world food supplies: The new frontiers. CHEMRAWN II. Perspectives and recommendations. Int. Rice Res. Inst., Los Baños, Philippines.

Swift, M.J. 1976. Species diversity and the structure of microbial communities. p. 185–222. *In* J.M. Anderson and A. Macfadyen (ed.) The role of terrestrial and aquatic organisms in decomposition processes. Blackwell Sci. Publ., Oxford.

Swift, M.J. 1984. Soil biological processes and tropical soil fertility: A proposal for a collaborative programme of research. Biol. Int. Spec. Issue 5. Int. Union of Biol. Sci., Paris.

Swift, M.J., P.G.H. Frost, B.M. Campbell, J.C. Hatton, and K.B. Wilson. 1989. Nitrogen cycling in farming systems derived from savanna: Perspectives and challenges. p. 63–76. *In* M. Clarholm and L. Bergström (ed.) Ecology of arable land. Developments in Plant and Soil Sciences 39. Kluwer Acad. Publ., Dordrecht, Holland.

Thomas, J. 1987a. Distribution of *Termitomyces* Heim and other fungi in the nests and major workers of *Macrotermes bellicosus* (Smeatham) in Nigeria. Soil Biol. Biochem. 19:329–333.

Thomas, R.J. 1987b. Factors affecting the distribution and activity of fungi in the nests of Macrotermitinae (Isoptera). Soil Biol. Biochem. 19:343–349.

Tisdall, J.M., and J.M. Oades. 1982. Organic matter and water stable aggregates in soils. J. Soil Sci. 33:141–163.

Visser, S.A. 1985. Physiological action of humic substances on microbial cells. Soil Biol. Biochem. 17(4):457–462.

Watson, J.A.L. 1972. An old mound of the spinifex termite, *Nasutitermes trioidae* (Froggatt) (Isoptera: Termitidae). J. Austral. Entomol. Soc. 11:79–80.

Wielemaker, W.G. 1984. Soil formation by termites, a study in the Kisii area, Kenya. Doctoral thesis. Agric. Univ., Wageningen, Netherlands.

Williams, M.A.J. 1968. Termites and soil development near Brocks Creek, Northern Territory. Aust. J. Sci. 31:153–154.

Wood, T.G. 1978. Food and feeding habits of termites. p. 55–80. *In* M.V. Brian (ed.) Production ecology of ants and termites. Cambridge Univ. Press, Cambridge.

Wood, T.G. 1979. The termite (Isoptera) fauna of Malesian and other tropical rainforests. *In* A.G. Marshall (ed.) The abundance of animals in Malesian rainforests. Trans. 6th Aberdeen-Hull Symp. on Malesian Ecology. Aberdeen Univ., 1978.

Wood, T.G. 1988. Termites and the soil environment. Biol. Fert. Soils 6:228–236.

Wood, T.G., and K.E. Lee. 1971. Abundance of mounds and competition among colonies of some Australian termite species. Pedobiologia 11:341–366.